Bisphenol A Removal from Water and Wastewater

Magdalena ZIELIŃSKA
Irena WOJNOWSKA-BARYŁA
Agnieszka CYDZIK-KWIATKOWSKA

Bisphenol A Removal from Water and Wastewater

Springer

Magdalena ZIELIŃSKA
University of Warmia
and Mazury in Olsztyn
Olsztyn
Poland

Agnieszka CYDZIK-KWIATKOWSKA
University of Warmia
and Mazury in Olsztyn
Olsztyn
Poland

Irena WOJNOWSKA-BARYŁA
University of Warmia
and Mazury in Olsztyn
Olsztyn
Poland

ISBN 978-3-319-92359-8 ISBN 978-3-319-92361-1 (eBook)
https://doi.org/10.1007/978-3-319-92361-1

Library of Congress Control Number: 2018942517

Printed on acid-free paper

This Springer imprint is published by the registered company Springer International Publishing AG part of Springer Nature
The registered company address is: Gewerbestrasse 11, 6330 Cham, Switzerland

Contents

Chapter 1
Introduction

Bisphenol A (BPA) is probably one of the most prevalent synthetic xenoestrogens in the environment. The prevalence of BPA is due to the rapid growth of industrial applications that require BPA. As a result, by 2010 the global production of BPA had increased to about 5 million tons per year. Although the single largest source of BPA contamination is difficult to identify, BPA is always man-made.

Ninety-two percent of all BPA enters the environment in wastewater from various sources. The concentration of BPA in wastewater ranges from ng/L up to over a dozen mg/L in certain hazardous-waste landfill leachates in Japan. The residual BPA that is discharged to the aquatic environment will often contaminate sources of drinking water. BPA in biosolids from wastewater treatment plants is also a source of concern. Its concentration in these solids ranges from 100 to 10,000 ng/g, which means that biosolids from these treatment plants can be a source of BPA contamination in soils and groundwater if used as fertilizer.

Although much recent research has focused on the toxic effects of BPA and its removal from wastewater, there is a lack of information about BPA in current standards for municipal wastewater treatment plants, such as the Water Framework Directive (WFD, The Directive 2000/60/EC). Much progress has been made in removing BPA from wastewater in recent years. Thus, this monograph aims to summarize up-to-date information about BPA and developments in the technology of wastewater treatment for the removal of micropollutants, using BPA as an example. The content of this book covers the sources and properties of BPA, physical and chemical pre-treatment and treatment technologies, as well as BPA biodegradation or metabolism in biological treatment technologies. The text also examines systems that integrate these techniques to remove BPA from wastewater.

Recent developments offer physical and chemical technologies for pre-treatment and treatment in BPA removal. These include adsorption, membrane filtration and a number of different options that are currently available for oxidation of BPA. All of these can be effective steps in the removal process. Although there is still a question whether the use of these technologies is feasible for full-scale plants due to cost, their effectiveness is high. Chemical pre-treatment is often used to partially oxidize

M. ZIELIŃSKA et al., *Bisphenol A Removal from Water and Wastewater*,
https://doi.org/10.1007/978-3-319-92361-1_1

BPA to biodegradable intermediates. In addition, sonochemical reactions have also been described as effective in degradation of organic compounds in water solutions.

Microorganisms play an important role in BPA removal. The molecule can be readily metabolized by many bacterial communities, and BPA-degrading bacterial strains have been isolated from various environments and identified. The first isolated bacterial strain (*Sphingomonas* sp. strain MV1) was from the sludge of a wastewater treatment plant. Such bacterial strains are capable of growing on BPA, using it as a sole source of carbon and energy. The metabolites produced during degradation of BPA under aerobic conditions have been exhaustively studied, and several BPA degradation pathways have been proposed. Some information is also available on the enzymes and genes that are involved in BPA degradation. The bioreactors with activated sludge, biofilm and aerobic granular biomass have been developed to enhance the BPA removal in biodegradation and/or sorption. Particularly, membrane bioreactors that support high concentrations of micro-organisms and long sludge age and provide an effluent free of suspended solids are of major interest for obtaining highly efficient BPA removal.

Furthermore, by combining biological removal of BPA with the physical and chemical technologies mentioned above, wastewater treatment efficiency can be enhanced. For example, methods based on heterogeneous photocatalysis, which relies on the generation of hydroxyl radicals, have been employed for the removal of BPA from wastewater. Research has also focused on advanced oxidation techniques such as UV/Fenton, UV/O_3, UV/H_2O_2, O_3/H_2O_2 and O_3/UV/TiO_2 treatments. Oxidation products and other byproducts created by physical and chemical pre-treatment can be metabolized by microorganisms. Oxidation can be used not only as pre-treatment, but also for degrading recalcitrant contaminants after the highly biodegradable part of wastewater has been eliminated. Thus, combining oxidation and biological treatments for wastewater decontamination is promising.

In addition, the persistence of BPA in the microbial biomass or sludge is a potential source of environmental contamination when the sludge is used as a soil amendment or fertilizer. This makes it necessary to understand the efficacy of different pre-treatment methods for removing BPA and its intermediate by-products from sludge. The information on the BPA content in the waste sludge, BPA removal in various sludge treatment methods and combined methods to diminish BPA concentration in the sludge, which affect the toxicity of intermediate by-products and the stability of the sludge in the environment, is given in this monograph.

This collection of state-of-the-art research will provide a vital and comprehensive reference for students and researchers, as well as wastewater treatment plant designers and operators who seek to protect humans and the environment from the harmful effects of BPA.

Chapter 2
Sources and Properties of BPA

BPA was first synthesized by Aleksandr P. Dianin in 1891 and was investigated for potential commercial use in the 1930s during a search for synthetic estrogens. In the 1940s and 50s, uses for BPA were identified in the plastics industry. The sources of BPA pollution in the environment and human surroundings include not only accidental spills and industrial wastes and effluents, but also many diffuse sources, including many common and household products. BPA concentrations are generally lower than 1 μg/L in aquatic environments; however, they are often two to three orders of magnitude higher in industrial effluents. Exposure to BPA has been associated with various negative health effects, including effects on reproduction and development, although its low-dose toxicity is still controversial. At this time, it is unknown which sources of BPA exposure contribute most to total exposure levels.

2.1 Production and Use

BPA is a chemical compound used in massive amounts in the production of synthetic polymers. Epoxy resins using BPA were first synthesized by chemists in the United States and Switzerland in the early 1950s. BPA is used in industry as an important intermediate in the production of the following resins and polymers: polycarbonate (PC), epoxy, polysulphone (PSU), polyacrylate, polyetherimide (PEI), unsaturated polyester and phenolic substances. BPA gives desirable characteristics to these products, and thus is widely used in many products. The use of BPA can be divided into two categories: one where the chemical structure of BPA is modified or polymerized, and another where BPA is used as an additive (Geens et al. 2011). The largest amount of BPA is polymerized to form polycarbonate (PC) and epoxy resin. Polycarbonate is a thermoplastic polymer, commonly synthesized by the condensation of BPA and phosgene. Because of its advantageous properties, PC is used in optical media, construction materials, and the

M. ZIELIŃSKA et al., *Bisphenol A Removal from Water and Wastewater*,
https://doi.org/10.1007/978-3-319-92361-1_2

electrical and electronics industries. A small fraction of PC is used for bottles and other food-contact applications (3%) and for medical and healthcare applications (3%). Thus, all these common products are potential sources of BPA contamination. The most common epoxy resins are formed by a reaction between epichlorhydrine and BPA. Because of the desirable properties of epoxy resins, they are used as coatings for consumer and industrial applications, protective coatings for automotive and marine uses, electrical and electronic laminates, adhesives, and paving applications. Internal coatings of coils and of food and beverage cans may contribute approximately 11% of epoxy resin usage (US EPA 1998b).

Polysulphone (PSU) is a thermoplastic polymer synthesized by the condensation of BPA and bis(4-chlorophenyl)sulphone. PSU can be used as an alternative to PC that is highly transparent, sterilizable, long-term dishwasher safe and impact resistant. The resistance of PSU to high temperatures also makes it a useful flame retardant without compromising its strength. Phosphorus-modified PSU and aromatic PSU can also be used as flame retardants (Perez et al. 2007). Dental composite resins consist of a mixture of monomers and are most commonly based on bisphenol-A glycidyl methacrylate (bis-GMA). In addition to bis-GMA, these resins contain other monomers to modify their properties, e.g. bisphenol-A dimethacrylate (bis-DMA) (Fung et al. 2000).

Polyetherimide (PEI) is used for non-burning, low-smoke-emission applications in the medical, electronic, electrical, automotive and aircraft industries, and in microwave applications. BPA is used to produce flame retardant tetrabromobisphenol-A (TBBPA). TBBPA contains trace amounts of BPA (typically less than 3 ppm). Around 18% of TBBPA is used for the production of TBBPA-derivatives and oligomers, and another 18% is used as an additive flame retardant. The presence of BPA in these products still needs further investigation (Covaci et al. 2009).

Polyester resins based on BPA include BPA fumarates and BPA dimethacrylates. Propoxylated BPA-fumarate unsaturated-polyester resin is resistant to a broad range of aqueous inorganic acids, salts and alkaline solutions. Thus, it is used in many applications involving highly corrosive environments such as storage tanks and process vessels. Glass reinforced composite resins are used to make strong, light laminates in the manufacture of boats, swimming pools, spa baths and translucent roof-sheets. BPA is also used as precursor in the synthesis of benzoxazines monomers. Polybenzoxazines are a new type of addition-cure phenolic system. They have gained interest because of their capability to exhibit the thermal and flame retardant properties of phenolics along with mechanical performance and molecular design flexibility. Therefore they can be used in a variety of applications in the industry of composites, coatings, adhesives and encapsulant's manufacturing (Ghosh et al. 2007).

BPA is widely used in the paper industry as a phenolic developer. In this application, BPA is present in its free, unpolymerized form and is thus easily available for uptake. Among other applications, thermal paper is used in point-of-sale receipts, prescription labels, airline tickets, and lottery tickets (Environmental Working Group 2010).

Because of its antioxidant properties, BPA is used as an additive in soft polyvinylchloride (PVC) and also as an inhibitor of vinyl-chloride polymerization in the production of PVC. To a lesser extent, BPA is used as an antioxidant in hydraulic brake fluids, and in tire manufacturing during the compounding phase and the curing process (European Union 2010).

BPA can be found in a wide variety of materials and products that people come into contact with on a daily basis (e.g. bottles, coatings, pipes, dental sealants, food packaging, nail polishes and flame-retardant materials). Very recently, the European Union banned the use of BPA in plastic infant-feeding bottles (Commission Directive 2011/8/EU 2011).

Any residual, unreacted BPA that remains in polycarbonate products and epoxy resins can leach out into food or the environment. Polycarbonate is generally stable, but some BPA can be released from polycarbonate when it is exposed to strongly basic conditions, UV light or high heat. Epoxy resins made with BPA are stable; only residual BPA is expected to be released from them. Alternatives to BPA exist, but there is no one replacement solution for all industrial applications (Flint et al. 2012). Switching to alternative chemicals involves trade-offs, and investment in new equipment may be necessary. Potential replacements for BPA-containing polycarbonates include acrylic, polyester, and polypropylene, but these materials have drawbacks: acrylic is not as strong and can yellow over time, polyester is more expensive, and polypropylene is not stable at high temperatures. Alternatives to BPA in can liners include polyester, polyacrylate, synthetic resins, and PVC pastes. Some of these alternatives also have detrimental effects. Other options exist, which have already been in use for decades and are easily recycled, such as glass, stainless steel, and aluminum. Replacement plastics may include high-density polyethylene, polyethylene terephthalate, and Grilamid TR-90. Polyester and oleoresins (plant derived) can also be used as replacement in can liners. Tetrapaks, constructed from composite paper, polyethylene, and aluminum foil, provide another packaging alternative.

2.2 Physical, Chemical and Biological Properties

BPA (Chemical Abstracts Service [CAS] number 80-05-7) is manufactured by condensation of 2 mol phenol with 1 mol acetone in the presence of an acid catalyst such as hydrogen chloride (O'Neil 2006). BPA is completely soluble in organic solvents and partially soluble in water. It exists at room temperature in the form of a white solid flake or crystal. The purity of BPA is 99–99.8% depending upon the manufacturer (Directive 2003/89/EC). Impurities typically include phenol (<0.06%), *ortho*- and *para*-isomers of BPA (<0.2%) and water (<0.2%). All of the identified components contained phenol groups. These impurities can increase the environmental risk posed by BPA.

The physical, chemical and biological properties of BPA provide basic information on its characteristics, give information necessary to describe the potential for environmental release of BPA, exposure to BPA, and its partitioning between media, and give insight into its potential to cause adverse toxicological effects (Table 2.1).

The water solubility of BPA provides an indication of its distribution between environmental media, the potential for environmental exposure through its release to aquatic compartments, and potential for human exposure through ingestion of drinking water. BPA has a moderately high water solubility range from 120 to 300 mg/L (about 200 mg/L at 25 °C). The solubility of BPA is a cause for concern, because it is higher than 1×10^{-5} g/L, and substances with a solubility above this level have more adverse effects and greater risk of potential aquatic and general population exposure due to their high bioavailability. The dissociation constant (pK_a) determines if a chemical will ionize under environmental conditions. This constant provides the amount of the dissociated and undissociated forms in the water. BPA has a pK_a between 9.59 and 11.30 (Staples et al. 1998) and therefore will not appreciably ionize at environmental pH levels that are generally 7 and lower.

The Henry's law constant provides an indication of the volatility of BPA in water. In environmental assessments, the Henry's law constant is typically determined for a substance in water at 25 °C. For BPA, this constant was calculated as 4.03×10^{-6} Pa m^3/mol at 25 °C using a vapour pressure of 5.3×10^{-6} Pa at 25 °C and a water solubility of 120–300 mg/L at 25 °C (HSDB 2000). This value of the

Table 2.1 Chemical and physical properties of BPA

Properties	Value
Formula	$C_{15}H_{16}O_2$ or $(CH_3)_2C(C_6H_4OH)_2$
Synonyms	Bis(4-hydroxyphenyl)dimethyl methane; 4,4′-dihydroxydiphenyl propane; 4,4′-dihydroxy-2,2-diphenyl propane; diphenylolpropane; 4,4′-isopropylidenediphenol
Boiling point	220 °C (at 4 mm Hg)
Melting point	150–155 °C (solidification range)
Specific gravity	1.195 at (25/25 °C)
Octanol/water partition coefficient	$\log K_{ow} = 3.32$
Water solubility	<1 mg/mL at 21.5 °C
Vapor pressure	3.91×10^{-7} mm Hg at 25 °C 0.2 mm Hg at 170 °C 1×10^{-8} mm Hg at 25 °C (estimated from Henry coefficient and water solubility)
Henry's law constant	1×10^{-6} to 1×10^{-5} Pa m^3/mol
Color/form	White crystals or flakes
Odor	Mild phenolic odor
Decomposition temperature	>200 °C

Henry's law constant suggests that only insignificant amounts of BPA will volatilize from all bodies of water. BPA is also not expected to volatilise significantly from wet or dry soil surfaces, due to its relatively low vapour pressure (5.32×10^{-5} Pa at 25 °C) and its tendency to adsorb to soil. However, because the vapour pressure increases considerably as temperature rises, potentially higher atmospheric concentrations of BPA are possible in certain conditions, such as during production or processing. Overall, the low volatility of BPA indicates that it is most likely to enter the environment as dust particles during production, processing or final use of products made with BPA. BPA will not often be present in its gaseous form in the environment because of its low vapor pressure.

Apart from the Henry's law constant, air-water partitioning coefficient (K_{aw}) indicates that volatilization is unlikely to remove a significant amount of BPA from water systems. To calculate the dimensionless K_{aw}, the formula H/RT is used, where H is the Henry's law constant (Pa m^3/mol), R is the gas constant [Pa m^3/(mol K)], and T is the absolute temperature (298 K). For BPA, K_{aw} is 1.7×10^{-9} m^3/m^3, which is a very low value. Chemicals with Henry's Law constants between 10^{-6} and 10^{-8} tend to partition predominately into water, whereas chemicals with a constant in the range 10^{-3} to 10^{-5} tend to distribute equally between liquid and gas phases (Eisenreich et al. 1981). An aqueous solution of BPA at a concentration of 16 ng/L will cause a corresponding concentration of BPA in the air of about 16×10^{-9} ng/L or 16 ng/m^3. In the atmosphere, BPA is a relatively short-lived compound.

In conclusion, BPA may have a tendency to partition into water, and its rate of evaporation from soil and water will be low, due to its moderately high water solubility (120 mg/L at 25 °C), its very low vapour pressure (5.32×10^{-6} Pa at 25 °C) and its low Henry's Law constant from 1.0×10^{-6} to 1.0×10^{-5} Pa m^3/mol.

One of the most useful properties for performing a hazard assessment of a substance is the octanol/water partition coefficient (K_{oc}), commonly expressed as its log value (i.e. log K_{ow}). The log K_{ow} of BPA ranges from 2.2 to 3.84. Because a log K_{ow} less than 1 indicates hydrophilic properties, and a log K_{ow} greater than 4 reflects hydrophobic properties, BPA does not have a strong affinity for either phase, but is instead found in various liquid and solid environmental matrices. In contrast, BPA in gaseous form will adsorb strongly to solid surfaces, including soils, vegetation, and aerosols, because its octanol-air partition coefficient is 2.6×10^{12}.

The extent to which BPA will adsorb onto soil, sediment or suspended solids depends on the organic matter, e.g. organic carbon content of these media. For risk assessment, K_{oc} can be derived from the following equation:

$$\log K_{oc} = 0.81 \quad \log K_{ow} = 0.10 \quad \text{for } 1.0 < \log K_{ow} < 7.5$$

The strength with which BPA is adsorbed to solids in the environment is important because it gives an indication of the compounds mobility. However, estimates of the mobility of BPA differ depending on the method used to calculate the strength of its adsorption to solids. Soil adsorption experiments indicated that the K_{oc} coefficients were between 636 and 931, classifying BPA as having low mobility (Fent et al. 2003). Dorn et al. (1987) estimated the K_{oc} values between 314 and 1524 suggesting that mobility of BPA in soil would be moderate to extensive. On the basis of the K_{oc} of 293–1524 and the moderately high log K_{ow} of BPA it is concluded that this chemical has a moderate potential for adsorption to organic materials. The best solution is to use the classification proposed by the U.S. Environmental Protection Agency (US EPA). The adsorption to soil within the context of the assessment is described qualitatively as very strong (above 30,000), strong (above 3000), moderate (above 300), low (above 30), and negligible (above 3). The higher K_{oc} value indicates stronger adsorption to soil. Strong adsorption may affect other processes, such as the rate of biodegradation, and may make the compound less bioavailable. Thus, based on the EPA classification, BPA can be considered to be moderately mobile.

With a standard fractional organic carbon content of soil, sediment and suspended sediment taken as 2, 5 and 10%, respectively, specific adsorption constants (K_p) for soil, sediment and suspended sediment can be calculated directly from the K_{oc} or K_{ow} value.

Partition coefficients for BPA calculated using a log K_{ow} of 3.40 are:

K_{oc}	715 L/kg	Organic carbon-water partition coefficient
K_{psoil}	14.3 L/kg	Solids-water partition coefficient in soil
K_{psed}	35.8 L/kg	Solids-water partition coefficient in sediment
K_{psusp}	71.5 L/kg	Solids-water partition coefficient in suspended matter
$K_{susp\text{-}water}$	18.8 m^3/m^3	Suspended matter-water partition coefficient
$K_{soil\text{-}water}$	21.7 m^3/m^3	Soil-water partition coefficient
$K_{sed\text{-}water}$	18.7 m^3/m^3	Sediment-water partition coefficient

The environmental distribution of BPA has been estimated to be 32% in sediments, 43% in water, 24% in soil and 3.5×10^{-5} % in air (Hanze 1994).

The persistence of BPA in the environment is determined by the type and rate of potential removal processes. These processes include oxidation and photolysis, which produces phenol, 4-isopropylphenol and a semiquinone derivative of BPA (Peltonen et al. 1986). In water, BPA is transformed by photolysis at wavelengths above 290 nm; this takes place most readily under alkaline conditions. However, photolysis is unlikely to remove a substantial amount of BPA because the compound binds to organic materials and may undergo sedimentation in aquatic systems. In the atmosphere, BPA exists almost entirely in the particulate phase. Although the compound can be removed from the atmosphere by photolysis, most atmospheric BPA is removed by oxidation by hydroxyl radicals and ozone. These hydroxyl radicals have an estimated half-life of 3.48 d^{-1}.Thus, BPA released to the atmosphere is likely to be degraded by reaction with hydroxyl radicals.

For determining the persistence of BPA, the importance of both aerobic and anaerobic biodegradation, as well as partitioning and transport in the environment, are considered to determine which removal processes are most likely to occur. Although BPA tends to be slowly biodegraded in water, soil and sediments, biodegradation plays a major role in elimination of BPA in the environment, gradually reducing the potential risk posed by the compound. Two factors which seem to be important in BPA biodegradation are the composition of the microbial community in the environment and the prevailing redox conditions. With acclimatization, the composition of the community will change, and the ability of individual species to use BPA as a carbon source will improve. With regard to community composition, certain species from several bacterial genera have been isolated that are capable of using BPA as a sole source of carbon and energy. The degradation efficiency of these BPA-degrading strains can be improved by another non-BPA-degrading strain. For example, a period of about 40 days was necessary for complete BPA removal by a BPA-degrading strain alone, whereas a mixture of two strains completely degraded BPA within 7 days (Sakai et al. 2007). With regard to acclimatization in general, experiments have found that the half-life of BPA with a non-acclimated community was 24 h to 6 months, or 96 h to 24 months, whereas that with an acclimated community was 2.5 to 4 days (Hanze 1994). The redox potential in the environment also has an effect because it determines which electron acceptors are used during anaerobic biodegradation.

The ability of a BPA to accumulate in living organisms is described by the bioconcentration, bioaccumulation, biomagnification, and/or trophic magnification factors.

Bioconcentration is considered to be a partitioning process between water and organisms and comparable with such processes as sorption and octanol-water partitioning. The potential for a molecule to be absorbed by an organism tends to be lower when the molecule is larger than 1000 Da. For the 1000 Da size to be an absolute threshold for absorption, biological systems are dynamic and even relatively large chemicals may be absorbed under certain conditions.

The bioconcentration and bioaccumulation potential of BPA is often compared to the K_{ow} coefficient. K_{ow} represents the lipophilicity and the hydrophobicity of a chemical and how it thermodynamically distributes, i.e., partitions, between aqueous and organic phases. K_{ow} is generally considered to be a reasonable substitute phase for lipids in biological organisms. The two physical-chemical properties K_{ow} and aqueous solubility (*SW*) are inversely related and uncertainty of measured and estimated values of K_{ow} generally increases for very hydrophobic chemicals, i.e. log K_{ow} values greater than about 6.

Bioconcentration of BPA in aquatic organisms can occur through uptake directly from the water (bioaccumulation) or through food (biomagnification). Bioaccumulation is a process in which BPA is absorbed in an organism by all routes of exposure possible in the natural environment, i.e. dietary and ambient environment sources. Organisms can be exposed to water with BPA until equilibrium is attained (internal contents do not increase anymore).

Bioaccumulation is the net result of the competing processes; this includes uptake, metabolism and elimination of a BPA in an organism. Bioaccumulation can be evaluated using the bioaccumulation factor (BAF), the steady state ratio of a chemical in an organism relative to its concentration in the ambient environment, where the organism is exposed through ingestion and direct contact. The experimental bioconcentration factors (BCFs) are more commonly used to evaluate the bioaccumulation hazard. BCFs are defined as the ratio of the concentration of a dissolved chemical in an aquatic organism ($C_{organism}$) to the concentration of the chemical in the exposure medium (C_{water}). The BCF can be calculated from the ratio between the content in organisms and water:

$$\mathrm{BCF} = C_{organism}/C_{water}$$

The BCF can be expressed on a lipid basis as well as on a fresh weight basis. According to commonly used guidelines for the classification of the environmental hazard of chemicals within the EU, a chemical with a log K_{ow} value >3 and/or a BCF value of >100 is considered to have a bioaccumulative potential (Nordic Council of Ministers 1996). With their high bioaccumulation potential and sensitivity to pollutants, fish have been recommended as an indicator of contaminated aquatic environments and a biomarker of the estrogenic potency of pollutants (van der Oost et al. 2003). The BCF for BPA is calculated using bioaccumulation by fish from the ambient environment only through its respiratory (e.g. gills in fish) and dermal surfaces, i.e. chemical exposure in the diet is not included (Arnot and Gobas 2006).

The biota-sediment accumulation factor (BSAF) is a simple index for predicting the bioaccumulation of hydrophobic organic compounds (HOCs) in fish and other aquatic biota. It uses measured concentrations in sediment based on equilibrium partitioning between the sediment organic carbon and biotic lipid pools (Wong et al. 2001). The BSAF model has been suggested as a first-level screening tool for predicting bioaccumulation potential under in situ river conditions, with a BSAF of 4 serving as a screening level for neutral HOCs (Boese et al. 1995). The BSAF for BPA was calculated by the following equation: BSAF = (BPA in fish/lipid content in fish)/(BPA in sediment/organic carbon in sediment). When taking the multiple exposure routes of two compounds and the possible additive effects of pollutant mixtures into consideration, nonylphenol (NP) and BPA might pose significant risks to human health (Lee et al. 2015).

In biomagnification, the activity of BPA in an organism depends from diet. The food web magnification factor (FWMF) represents the average increase or decrease in lipid-normalized chemical concentrations for a unit increase in trophic position (Mackintosh et al. 2004). A FWMF greater than 1 indicates an occurrence of biomagnification in the food web, whereas a value less than 1 indicates trophic dilution. The trophic magnification factor (TMF) is analogous to the FWMF and is also used to identify food web biomagnification (Tomy et al. 2004). Biomagnification is expressed by a biomagnification factor (BMF), defined as the ratio of the chemical concentration in an organism (CB) to that in its diet (CD) at steady state, i.e. BMF = CB/CD. These concentrations can be expressed on a wet weight basis or dry weight basis, i.e. BMFWW or BMFDW. For lipophilic substances this can be achieved by expressing BPA concentrations in the organism and its diet on a lipid normalized or lipid weight (LW) basis, i.e. BMFLW = *C*B (LW)/*C*D(LW) (Arnot and Gobas 2006). Due to the metabolizable nature of BPA (Takeuchi et al. 2009), it is not biomagnified up the food web to higher trophic levels. Thus, BPA in plastic could be a significant exposure pathway of these chemicals to the higher-trophic-level animals.

The risk quotient (RQ) approach was used to characterize the ecotoxicological risks posed by BPA in rivers based on the US EPA guidelines (US EPA 1998a). The RQ was calculated as follows: RQ = MEC/PNEC, where MEC is the measured environmental concentration of BPA, PNEC is the predicted no-effect concentration. The MECs were determined by Lee et al. (2013). The PNECs of 0.06 μg/L for BPA in freshwater environment were determined by Wright-Walters et al. (2011).

The non-cancer hazard quotient (HQ) (US EPA 1989) is used to characterize the health risk posed by these compounds as follows:

$$HQ = ADD/RfD$$
$$ADD = (CF \cdot IR)/BW$$

where ADD is the average daily uptake of a chemical (μg/kg BW/day) and RfD is the daily intake reference dose (μg/kg BW/day). CF is the chemical concentration in fish (μg/kg ww). IR is the fish consumption rate (kg/person/day) for human populations of different age groups. BW is the body weight (kg) of populations of different age groups. An HQ value >1 indicates that there may be concern for potential human health effects, and vice versa.

There is controversy about whether effects seen at lower doses in animals (less than 1 mg/kg/day) are meaningful and relevant to humans. These low-dose effects are endocrine-related and include effects on puberty and developmental neurotoxicological effects (brain, behavior) at doses in animal studies as low as 2 μg/kg BW/day. Early investigations showed low toxicity of BPA, its fast metabolism and inconsiderable contamination of food with this compound, which determined the usage of BPA for the synthesis of polymers utilized in the production of food and water containers. In 1988, US EPA introduced the first safety standards for the use of BPA and its potential contact with food. Moreover, US EPA estimated the reference dose, which is 50 μg BPA/kg BW/day (Vogel 2009). The above dose remains an actual safety standard of the exposure of human organisms to BPA. This is the tolerable daily intake (TDI), as established by the European Food Safety Authority (EFSA 2010). Nevertheless, the abundance of new data concerning BPA toxicity and the fact that human organisms are more susceptible than rodents (model mammals used in the investigations of BPA toxicity) to deleterious action of this substance show the necessity of estimation of renewed safety dose for BPA (Rubin 2011).

Even though the toxicity of BPA has been quite extensively investigated compared to many other compounds, there is still no consensus regarding at what exposure levels BPA poses a human health risk. Low-dose effects, which have been increasingly reported during the last decade, include developmental effects on the brain, behavior and reproductive tissues in rodents. Such effects have been reported at doses in the range of a few micrograms per kilogram body weight per day, which are approximately 1000 times lower than the dose levels currently accepted by regulatory authorities as the No Observed Adverse Effect Level (NOAEL) and on which TDI in the US and Europe are based, i.e. 5 mg/kg BW/d. A low-dose effect was determined based on the US EPA criterion for low-dose effects of endocrine disrupting compounds at concentrations below those used for traditional toxicological studies. Levels below the current Lowest Observed Adverse Effect Level (LOAEL) of 50 μg/kg BW/day were considered as low dose for in vivo studies. For in vitro cell or organ culture studies, estimates of circulating levels of BPA at the LOAEL cut-off have defined an equivalent low dose concentration as <50 ng/mL or $<2.19 \times 10^{-7}$ M (Welshons et al. 2006).

There is general agreement that BPA is a reproductive and developmental toxicant at doses in animal studies of >50 mg/kg BW/day (delayed puberty in male and female rats and male mice), >235 mg/kg BW/day (reduced fetal or birth weight or growth early in life, effects on testis of male rats) and >500 mg/kg BW/day (possible decreased fertility in mice, altered estrous cycling in female rats). Systemic effects (reduction in body weight, changes in relative organ weights and increases in liver toxicity) were observed at doses above 5 mg/kg BW/day (identified as a NOAEL; LOAEL of 50 mg/kg BW/day).

The intermolecular interactions of BPA with biomacromolecules including proteins, enzymes and DNA are crucial to understand their toxicity mechanisms. Among these biomacromolecules, DNA is a genetic material indicating important biological roles in gene expression, transcription, mutagenesis, and carcinogenesis. DNA offers several points of close contact with the surface of BPA preferred to intercalate into the double helix of DNA and to sit in the minor groove of DNA (Wang et al. 2014). There are three binding modes including electrostatic binding, groove binding and intercalative binding in which molecules can be non-covalently attached to DNA. Molecular structure of BPA has aromatic rings and two –OH groups, which could play different roles in their binding with DNA. The aromatic rings system could make the rings enter into the narrower regions of DNA groove; however, two –OH groups could form van der Waals and/or hydrogen bond interactions with the helix of DNA. The hydrogen bond interactions existing between –OH groups of BPA and DNA could play an important role in the binding DNA due to the substitution of one $-CH_3$ of BPA. As for BPA, two benzene rings with –OH face toward the interface of DNA. The π-stacking interactions between the aromatic rings of BPA and the nucleic bases of DNA may be involved in the binding processes (Wang and Zhang 2015).

The US EPA has defined an environmental endocrine disruptor or endocrine disrupting chemical (EDC) as an exogenous agent that interferes with the production, release, transport, metabolism, binding, action, or elimination of natural hormones in the body responsible for the maintenance of homeostasis and the regulation of developmental processes. This definition is not limited to endocrine disrupting effects exclusive of the estrogen system. Rather, endocrine disruption encompasses effects on other endocrine systems including effects mediated by androgens, thyroid hormone, prolactin and insulin, among others. BPA is an endocrine disrupting compound in the broad sense of the definition. The pleiotropic mechanisms of BPA action causes that it is not simply a selective estrogen receptor modulator that binds nuclear estrogen receptors and acts as estrogen agonists in some tissues and estrogen antagonists in other tissues, or exclusively as an environmental estrogen (Wetherill et al. 2007).

BPA mainly affects the offspring viability, sex differentiation, immune hypersensitivity and gender differentiated morphology of the endocrine system when prenatally exposed (Golub et al. 2010). The range of pathways with which BPA potentially interferes may be much wider than expected. Additive, synergistic or antagonistic effects by simultaneous exposure to other compounds may occur.

BPA disturbs function of endocrine system (Vandenberg et al. 2007, Flint et al. 2012). It was proven that BPA behaves similarly to natural estrogen—17β-estradiol. BPA induced estrogen receptors but in the concentrations about one thousand higher (10^{-6} to 10^{-4} M) in comparison to estradiol. BPA was qualified as weak environmental estrogen, which activity towards classical nuclear steroid receptors ER_{β} and ER_{α} was over 1000–10,000 lower in comparison to 17β-estradiol. BPA even in very low (pico- and nanomolar) concentrations exerted multidirectional effects on physiological functions of cells and tissues by binding with receptors present out of the nucleus (Wetherill et al. 2007). The activity of BPA was estimated to be 2×10^{-3} lower than that of estradiol (Fürhacker et al. 2000).

It is well established that BPA can exert some of its effects by binding at the ER_{β} and ER_{α} receptors to induce estrogenic signals that modify estrogen-responsive gene expression. Mechanisms for ER-mediated gene regulation are complex and depend on the recruitment of tissue-specific co-regulatory factors that differentially affect the interaction of ER_{β} with ERE_{α} of different target genes. BPA selectively binds to ER_{α} and ER_{β} and has a higher affinity for ER_{β} in target cells. The binding affinity relative to 17β-estradiol for BPA also showed that BPA could bind to human ER_{α} and reported an IC50 (the concentration of chemical required to reduce specific 17β-estradiol binding by 50%) of 7.8 μM (Satoh et al. 2004). As an endocrine disruptor, BPA is not merely a weak estrogen mimic, but it exhibits characteristics of a distinct molecular mechanism of action at ER_{α}. The interactions of BPA and 17β-estradiol with an estrogen receptor can induce differential molecular effects, thus ultimately impacting the physiological response of sensitive cells. Estrogenic activity is estimated in terms of estradiol equivalent quantity (EEQ_i) by employing the following equations:

$$EEQ = C_i \cdot EEF_i \quad \text{and} \quad EEQ_t = \sum EEQ_i$$

where *i* refers to the compound *i* with concentration C, EEQ_t is the total estradiol equivalent quantity and EEF_i is the estradiol equivalency factor.

The EEFs were defined as the EC50 of each compound relative to the EC50 of 17β-estradiol, where EC50 is the contaminant concentration that produces a 50% maximal estrogenic response (Brix et al. 2010). Considering the daily water intakes estimated by US EPA (US EPA 1997) as 0.18 L/d for children and 1.4 L/d for adults, the levels of exposure via water ingestion in terms of EEQ were estimated to be 0.019 ng/d for children and 0.15 ng/d for adults.

The ubiquity of BPA products meant there are many potential sources of exposure to this synthetic estrogen. And yet, although BPA's estrogen-like properties (or estrogenicity) were never completely forgotten, its safety was defined by its commercial use in plastics and, accordingly, by its toxic rather than hormone-like properties. For the past 50 years, BPA's safety, along with that of most chemicals, has been defined according to the scientific presumption that the dose-response relationship is monotonic, that is, with increasing dose the effect increases and *vice versa*. Thus, at some diminished level of dose, the effect is marginal. Legally, this is called the minimis standard. Considering BPA to be a noncarcinogen, the US EPA used the lowest dose from the carcinogenesis study as the lowest observed adverse effect level and divided this number by an uncertainty factor of 1000 to determine a reference dose of 50 μg/kg BW/day (US EPA 2008). The 1000-fold uncertainty factor was the safety margin between the lowest observed.

There is not enough evidence whether BPA in the concentrations usually determined in the environment or/and in human organism (blood, urine) may affect humans. The results concerning neurotoxic and teratogenic effects of BPA are inconsistent; there is no strong evidence that BPA exhibits those activities in humans. It was found that biotransformation (oxidation) of BPA by cytochrome P450 monooxygenases in mammals usually enhances its toxicity and estrogenicity, which is associated with formation of reactive metabolites of BPA. The most of BPA metabolites formed during plant and microorganisms transformations exhibit lower toxicity and estrogenicity than BPA. There are indicated the following effects to be likely but requiring more evidence: (1) BPA may be associated with increased cancers of the hematopoietic system and significant increases in interstitial cell tumors of the testes, (2) BPA alters microtubule function and can induce an euploidy in some cells and tissues, (3) early life exposure to BPA may induce or predispose to preneoplastic lesions of the mammary gland and prostate gland in adult life, (4) prenatal exposure to diverse and environmentally relevant doses of BPA alters mammary gland development in mice, increasing endpoints that are considered markers of breast cancer risk in humans. Based on existing evidence, BPA may induce in vitro cellular transformation and, in advanced prostate cancers with androgen receptor mutations, may promote tumor progression and reduce time of recurrence (Wang et al. 2014).

2.3 Identification and Quantification of BPA in Groundwater, Surface Water, Wastewater and Leachate

BPA is a pseudo-persistent chemical, which despite its short half-life is ubiquitous in the environment because of continuous release. Release can occur during chemical manufacture, transport and processing. Post-consumer releases are primarily via effluent discharge from municipal wastewater treatment plants, leaching from landfills, combustion of domestic waste and the natural breakdown of plastics in the environment (US EPA 2011).

Rivers, lakes, and estuaries are major sinks for BPA. These surface waters accumulate BPA leached from plastic debris and landfill wastes along with BPA-containing sewage and effluent. BPA was one of the most relevant compounds detected in European ground waters, i.e. in terms of frequency of detection (40%) and maximum concentration levels (2.3 mg/L). The U.S. Geological Survey collected samples from 47 ambient groundwater sites in 18 States and analyzed them for 65 organic wastewater contaminants. BPA was detected in 29.8% of the sampled groundwater sites, with a mean detected concentration of 1.78 μg/L and a range of 1.06–2.55 μg/L, and was among the top five most frequently detected organic compounds (Barnes et al. 2008a, b). The Austrian groundwater study from the year 2000 has shown that the most abundant industrial chemicals found in ground water samples were BPA and NP with maximum concentrations of 930 ng/L and 1500 ng/L (Hohenblum et al. 2004). In ground water, BPA and NP levels were >1 mg/L in several places and were higher than the concentrations in river water (Reinstorf et al. 2008). A possible reason for this might be the efficient removal of BPA (>80%) during wastewater treatment (Clara et al. 2005). The ubiquitous presence of BPA in urban ground water results from a combination of local river water infiltration, sewer exfiltration, and urban stormwater recharge/runoff (Osenbrück et al. 2007). It appears that BPA is persistent under anaerobic conditions in groundwater (Ying et al. 2003).

The analysis of BPA concentrations in areas that were known or suspected to have at least some human and/or animal wastewater sources detected BPA in 9.5% of the samples at a reporting level of 1 μg/L. The maximum concentration of BPA measured in these samples was 1.9 μg/L (Barnes et al. 2008a; Focazio et al. 2008).

Most of environmental monitoring results show that the concentrations of BPA in surface water bodies are lower than 1 μg/L, mainly due to its partitioning and biodegradability (Tsai 2006). Current predicted no effect concentrations for organisms are 1.5 μg/L (EU), 1.6 μg/L (Japan) and 0.175 μg/L (Canada) (US EPA 2010). BPA was detected at a median concentration of 0.14 μg/L and a maximum concentration of 12 μg/L in 41.2% of 85 samples collected from the US streams in 1999 and 2000. The maximum concentration of 12 μg/L may be an outlier as it was much higher than any of the other samples (Kolpin et al. 2002). BPA monitoring studies found that out of 26 studies in North America (2 in Canada and 24 in the US), 80% (852 out of 1068) surface water samples had BPA concentrations below the detection limit. The reported median concentration was 0.081 μg/L and the 95th percentile concentration was 0.47 μg/L (Klečka et al. 2009). The results suggested that the frequency of locations in which BPA concentrations are likely to cause adverse effects on aquatic ecosystems is low. The concentrations of BPA observed in surface water vary considerably depending on the location and sampling period. Observed surface-water concentrations of BPA in the US range from 0.147 to 12 mg/L (Kolpin et al. 2002). BPA dissolved in surface water has a short half-life because of photo- and microbial degradation. Additionally, while most values reported for BPA in surface water are below 1 mg/L BPA, its concentrations can vary with depth (Funakoshi and Kasuya 2009) so sampling throughout the water column may be necessary to accurately characterize BPA concentration.

Conventional drinking water treatment plants are not specifically designed to remove EDCs and thus BPA may persist through treatment and be passed along to water utility customers. BPA in drinking water was reported to mainly stem from epoxy and polyester-styrene resins used in lacquer coatings of concrete tanks and the lining of steel pipes in water supply systems. The concentrations of BPA in surface and drinking waters were up to 8 ng/mL (Zhao et al. 2009).

In France, the Midi-Pyrenees DRASS (regional health and social affairs authority) and the Adour Garonne Water Agency conducted an analysis campaign in 2006–2007. BPA was screened for in 10 samples of raw groundwater, 31 samples of raw surface water and 40 samples of drinking water intended for human consumption. All results were below the method's detection level of 50 ng/L. All the data from the literature mention concentrations of BPA potentially present in drinking water intended for human consumption around a nanogram per litre. The detected BPA in surface waters (rivers) in Austria, Belgium, Switzerland, Germany and the Netherlands, at concentrations of tens or hundreds of ng/L (0.8 μg/L maximum) (Wenzel et al. 2003). In drinking water, the results were below the detection limit of 8 or 11 ng/L (one water supplier reported a value of 0.12 μg/L which was higher than the value in the corresponding raw water). The work of Kuch and Ballschmiter (2001) in Germany revealed BPA concentrations ranging from 500 pg/L to 16 ng/L in river water and from 300 pg/L to 2 ng/L in drinking water.

In the effluent from a WWTP, BPA was not detected above the detection limit of 0.0001 μg/L in samples collected from surface water or in drinking water at various stages of treatment at plants (Boyd et al. 2003). A California study detected BPA in two of three treated wastewater samples at 0.38 and 0.31 μg/L (limit of detection = 0.25 μg/L) (Jackson and Sutton 2008). BPA was detected in wastewater generated by a pharmaceutical manufacturer (0.295 μg/L), an industrial laundry (21.5 μg/L), and a paper products manufacturer (0.753 μg/L). A Canadian papers report BPA from 0.031 to 49.9 μg/L in sewage influent and effluent (generally <1 μg/L in the influent and <0.3 μg/L in the effluent) and from 0.104 to 36.7 μg/g in raw and digested sewage sludge from WWTPs (Lee and Peart 2000b). BPA was detected in all sewage samples from 31 WWTPs across Canada with concentrations ranging from 0.080 to 4.98 μg/L (median 0.329 μg/L) for the influent and from 0.010 to 1.08 μg/L (median 0.136 μg/L) for the effluent (Lee and Peart 2000a).

Based on comparison of influent and effluent levels, it was estimated that BPA in the influent was removed by the sewage treatment process with a median reduction rate of 68%. BPA was detected in sludge samples at concentrations ranging from 0.033 to 36.7 μg/g on a dry weight basis. A wide range of BPA was detected in wastewater discharges from industrial facilities with concentrations ranging from 0.23 to 149.2 μg/L. Higher BPA levels in wastewater were associated with facilities producing chemicals and chemical products and packaging and paper products, and with commercial dry cleaning establishments. BPA concentrations in pulp and paper mill sludge ranged from <0.02 (below detection limit) to 3.33 μg/g, with a median value of 0.076 μg/g, on a dry weight basis (Lee and Peart 2000a). The concentrations of BPA ranged from 0.088 to 11.8 μg/L in WWTP influent; the removal efficiency was about 71% (Melcer and Klečka 2011).

BPA in biosolids, products from WWTPs was found in concentrations between 1090 and 14,400 μg/kg BPA (median 4690 μg/kg) (Kinney et al. 2006). BPA in treated biosolids from a municipal WWTP was 4600 μg/kg and 81 μg/kg in soil that received the land applied biosolids, 147 μg/kg in a nearby control soil that did not receive treatment with biosolids (Kinney et al. 2008) and 81 μg/kg in earthworms living in treated soil (Staples et al. 2010).

BPA has been detected in landfill leachate with maximum concentrations of 1.7 and 1.4 μg/L in the receiving groundwater plume at a landfill that was known to be leaking (Rudel et al. 1998), and with concentration above 500 μg/L of BPA (Tsai 2006). At Japanese landfills, maximum concentrations of BPA in untreated leachate was 17,200 μg/L while after treatment the BPA concentration dropped to 5.1 μg/L (Crain et al. 2007). BPA concentrations in soil samples taken from outdoor play areas of homes and daycare centers ranged from 4 to 14 μg/kg dw, with means of 6–7 μg/kg dw (Wilson et al. 2007). A median concentration of BPA was 0.6 μg/kg in freshwater sediments, including samples with measurements below the detection limit; BPA concentrations in samples from the US ranged from 1.4 to 140 ppb dry weight (Klečka et al. 2009). The data indicate the prevalence of BPA in the environment and the need to use technologies in municipal industry that reduce landfilling of waste and increasing the degree of removal of BPA from wastewater. In addition, the dissemination of advanced methods of chemical oxidation technologies for water treatment is required.

During measurements of BPA concentrations in different media, there are many challenges associated with sample preparations. The recovery efficiency via the extraction and elution of compounds depends on a wide range of physico-chemical properties of these compounds. For example, polar compounds may be lost during solid-phase extraction (SPE) owing to their low affinity for hydrophobic sorbents, while non-polar compounds are retained but may be difficult to desorb them during elution. Non-polar compounds are also more likely to adsorb on flask walls and connecting tubing, leading to diminished analytes recoveries. In addition, the extraction of ionized compounds requires lowering sample pH which can lead to the co-extraction of undesired interferences such as natural organic matter (NOM). Although more time-consuming and labour-intensive, standard additions have been found to be the most reliable and efficient method to minimise matrix effects.

There are no standardized methods and no standard validation protocols available for the analysis of emerging micro-pollutants in environment. Directive 2006/11/CE (European Parliament and The Council 2006) establishes a list of compounds, including BPA, to be monitored in discharges and mentions some measures to protect the aquatic environment. The substances of concern are grouped in List I that includes dangerous and hazardous compounds to be eliminated from discharges, including sewers, and List II where dangerous substances should be diminished in discharges. Concerning this regulation, it is necessary to develop reliable and multi residual methodologies to analyze a BPA from different environment matrix.

There are various sample preparation techniques with common objectives: remove potential interferences from the sample matrix; if necessary, convert the analytes into a more suitable form; increase the concentration of the target analytes and provide a reproducible method that is independent of the sample matrix. Matrix effect is one of the main disadvantages associated with liquid chromatography-mass spectrometry (LC-MS), especially when working in the electro spray ionization mode. Common methods to reduce matrix effects are optimizing the step of samples clean up, minimizing the co-extraction and co-elution of matrix compounds such as NOM, and optimizing chromatographic separation. In spite of substantial technological advances in analytical field, most instruments cannot directly handle complex sample matrixes yet. As a result, sample-preparation steps are liquid-liquid extraction (LLE) or solid phase extraction (SPE).

LLE, based on the transfer of analyte from the aqueous sample to a water-immiscible solvent, is employed for sample preparation. LLE labour is intensive, expensive, time-consuming and environmentally unfriendly. Liquid-phase microextraction (LPME) is an alternative miniaturized sample-preparation approach. In LPME, a microliter volume of the solvent is needed to extract analytes from the aqueous samples. Single drop microextraction (SDME) was developed as a solvent-minimized sample pretreatment procedure. As a solution to improve the stability and reliability of LPME, Pedersen-Bjergaard and Rasmussen (1999) introduced hollow fiber liquid-phase microextraction (HF-LPME) that allows extraction and pre-concentration of analytes from complex samples in a simple and inexpensive way. Dispersive liquid–liquid microextraction (DLLME) was introduced by Ahmadi et al. (2006). It is a simple and fast microextraction technique based on the use of an appropriate extractant, i.e. a few microliters of an organic solvent such as chlorobenzene, chloroform or carbon disulfide with high density and a disperser solvent such as methanol, acetonitrile or acetone with high miscibility in both extractant and aqueous phases. For determining of BPA in water samples, the DLLME and high performance liquid chromatograph (HPLC) connected to a UV detector (HPLC-UV) method has been used (Rezaee et al. 2006). This method showed good linear range (0.5–100 μg/L) without using any derivatization reagent or applying very sensitive determination methods as gas chromatography mass spectrometry GC–MS and HPLC–MS. The main drawback of DLLME is the consumption of higher volumes (i.e. mL) of disperser solvent. Some progress has been made to use ultrasonic energy to disperse the extraction solvent in the absence of disperser solvent. The DLLME and heart-cutting multidimensional gas chromatography coupled to mass spectrometry was developed for the determination of free and total BPA and bisphenol B (BPB) in human urine samples (Cunha and Fernandes 2010). The efficiency of extraction (71–93%) and acceptable total DLLME yields (56–77%) were obtained for analyte. The limits of detection for BPA were 0.03 and 0.05 μg/L. Free BPA was detected in 45% of the urinary samples whereas total BPA was detected in 85% of the samples at concentrations ranging between 0.39 and 4.99 μg/L.

Other methods of sample preparation are microwave extraction (ME) and supercritical fluid extraction (SFE). One of the most frequently applied methods based on the phase boundary processes is static and dynamic headspace (HS) and purge and trap technique (PT). An alternative to HS and SPE is solid phase microextraction (SPME). In SPME, coated fibers are used to isolate and concentrate analytes into a range of coating materials. After extraction, the fibers are transferred, with the help of the syringe-like handling device, to analytical instruments for the separation and quantification of the target analytes. During the SPME, an equilibrium between the coated fiber and sample matrix is observed. At equilibrium, $k = C_s/C_a$, where C_s is the concentration of the organic compound in the solid phase, C_a is the concentration of BPA in aqueous phase, and k is the partition coefficient for solid phase/aqueous phase system. The k values have been calculated for many environmental compounds and defined efficiency of pre-concentration of BPA from water sample (Arthur and Pawliszyn 1990).

In SPME, a fused silica fiber is coated with a sorbent to extract sample components and pass analytes directly with different separation techniques such as capillary gas chromatography (CGC), GC/GCMS, LC/LC-MS, GC-ICP-MS, and HPLC-UV/HPLC-MS. To date, many materials, such as polydimethylsiloxane, Carbowax (polyethylene glycol), polyacrylate, polypyrrole and molecular imprinted polymer (MIP) have been used as SPME fibers coating. MIPs are polymeric materials that exhibit high binding capacity and selectivity against a target molecule that is purposely present during the synthesis process. In the common approach, the synthesis of MIPs first involves the solution complexation of a template molecule with functional monomers, followed by polymerization of these monomers around the template with the aid of a cross-linker in the presence of an initiator. After the removal of the template by chemical reaction or extraction, binding sites are exposed that are complementary to the template in size, shape and position of the functional groups, which consequently allow its selective uptake.

Chemical stability and easy preparation make MIP suitable as SPME fibers coating (MISPME – SPME based on MIP coating). A simple preparation approach for BPA, using MISPME was developed by Tan et al. (2009). A capillary was inserted into a larger bore capillary to form a sleeve as a mold. The prepolymer solution which was composed of BPA, acrylamide, 3-(trimethoxysilyl)propyl methacrylate, azobis(isobutyronitrile) and acetonitrile was introduced into the interspace between the two capillaries, followed by polymerization under UV photoirradiation. Watabe et al. (2004) prepared uniformly sized polymer particles for BPA trapping using *p-tert*-butylphenol as a pseudo-template through the two-step swelling and polymerization method. This MIP for BPA was combined with column-switching HPLC and applied to the actual determination of BPA at a level of ng/L in environmental water. MIP synthesis can be simple and cheap, and the resulting MIPs are unaffected by changes in heat and pH. MIPs typically display high cross-reactivity and are not particularly amenable to chemical modifications.

BPA can be oxidized and exhibits electrochemical activity since it contains phenolic hydroxyl groups. Electrochemical removal of BPA based on its oxidation was successfully fulfilled, and HPLC coupled electrochemical detection was used for the determination of BPA. Electrochemical sensors have great potential for environmental monitoring because of their portability, field deploy ability, excellent sensitivity (in low ppb levels), automation, short analysis time, low power consumption and inexpensive equipment. BPA is electrochemical active, direct determination of BPA using electrochemical sensor is rare because the response of BPA at traditional electrochemical sensor is very poor (Wang et al. 2009). To overcome these drawbacks, novel sensing material with high sensitivity and rapid response must be developed. Since the discovery of ordered mesoporous silica molecular sieves, the interest in this research field has expanded all over the world. With the characteristics such as highly uniform channels, large surface area, narrow pore-size distribution and tunable pore sizes over a wide range, mesoporous materials have attracted considerable attention and obtained wide applications in catalysis as well as in other realms of chemistry. Undoubtedly, these excellent properties will result in better charge transport and improved electrochemical signal, so mesoporous material is considered as a promising sensing material for electrochemical sensor. For instance, the mesoporous silica was successfully used to improve the electrochemical responses of compounds as well as biomolecules. The mesoporous silica sensor greatly enhances the response signal of BPA due to the large active surface area and high accumulation efficiency. The linear range is from 2.2×10^{-7} to 8.8×10^{-6} mol/L, and the limit of detection was 3.8×10^{-8} mol/L. Finally, the mesoporous silica sensor was successfully employed to determine BPA in water samples (Wang et al. 2009).

Electrochemical procedures are an alternative way to deposit coatings on SPME fibers (Li et al. 2016). The idea is to coat the metal fiber used in the SPME instrument with a conducting polymer, and cyclic voltammetry or potentiometry are usually used for this purpose. These techniques are inexpensive, and fiber coatings of the oxidized or reduced form of the polymer can also be obtained. The sensing materials include porous inorganic materials, electrocatalysts, MIPs, conducting polymers, carbon nanomaterials, transition-metal hybrids, metal oxide hybrids and enzymes that have inherent properties for detection of BPA, such as good adsorption/accumulation capacity, large surface area, good electrical conductivity, favourable electron-transferability and electro-catalytic properties. A major obstacle encountered in the detection of BPA by electrochemical method is the relatively high over potential together with poor reproducibility although BPA can be oxidized at electrode surface due to containing phenolic hydroxyl group. The bare electrode often suffers from a fouling effect, which causes rather poor selectivity and sensitivity. An effective way to overcome these barriers is electrode modification, of which is capable of reducing the overvoltage and overcoming the slow kinetics of many electrode processes. Owing to high electrical catalytic properties, high chemical stability and extremely high mechanical strength, multi-walled carbon nanotubes (MWCNTs) have been used to modify an electrode. The electrochemical sensor for the determination of BPA in water was fabricated by immobilization of MWCNTs/melamine complex onto the surface of glassy carbon electrode. This sensor exhibited strong catalytic activity toward the oxidation of BPA with a well-defined cyclic voltammetric peak at 0.56 V and exhibited a wider linearity range from 10.0 nM to 40.8 μM BPA with a detection limit of 5.0 nM (Li et al. 2012).

Some modified electrodes have been developed for investigating the electrochemical behaviour of BPA. The mesoporous carbon possesses large surface area, well defined pore size and pore volume, good conductivity, high thermal stability, flexible frame work composition and chemical inertness. It shows better performance on the adsorption of BPA (Li et al. 2016). Traditional carbon paste electrodes are composed of graphite powder sand and paraffin oil. The components are gradually replaced by new materials, for example, ionic liquids are used as binders instead of paraffin oil. Ionic liquids possess high ionic conductivity and high viscosity, and they are promising candidate materials for fabrication of electrochemical sensors. The developed ionic liquid-based carbon paste electrodes provide a fast electron transfer rate and a decrease in the over potentials of some organic compounds. The nanocarbon ionic liquid past electrode was composed of nano-graphite powders, ionic liquid and paraffin oil (Li et al. 2016).

The demand for rapid, sensitive and accurate methods to detect BPA in different environmental matrix has increased. In particular, tests that can be completed within minutes or hours would enable processors to take quick corrective actions when BPA is detected. The importance of biosensors relies on their high specificity and sensitivity, which allows the detection of a broad spectrum of analytes in complex samples with minimum sample pre-treatment. According to definition proposed by IUPAC (Thevenot et al. 1999), a biosensor is a self-contained integrated device which is capable of providing specific quantitative or semi-quantitative analytical information using a biological recognition element (biochemical receptor) which is in direct spatial contact with a transducer element. Biosensors involve biological recognition components such as enzymes, nucleic acids, antibodies, animal or vegetable tissues in intimate contact with an appropriate transducer. When antibodies or antibody fragments are used as molecular recognition element for specific analytes (antigens) to form a stable complex, the device is called immunosensor. Depending on the method of signal transduction, immunosensors may be divided into four basic groups: electrochemical, optical, piezoelectric and thermometric. The transducers chosen are directly related to the labelling, enzymatic or not, performed on the antigen or on the antibody. For each particular detection type, a specific labelling is usually performed, even though some labels can be used with different detection methods (i.e. horseradish peroxidase can be employed for an electrochemical immunosensor and for fluorescent and chemiluminescent detection using a fibre optic sensor). Most of the developed immunosensors are based either on a competitive or sandwich assay, when applied to the detection of low and high molecular weight molecules, respectively (Thevenot et al. 2001).

Two approaches can be followed for the development of competitive immunosensors: a first one in which immobilised antibodies (Ab) react with the free antigens (Ag) in competition with labelled antigens (Ag*), and a second one in which immobilised antigens compete with free antigens for labelled free antibodies (Ab*). To favor the immobilization and the interaction with the antibody, the antigens, especially those with a small molecular weight, are usually conjugated with a protein. Both these approaches are defined as direct competitive immunoassay. The second format is generally preferred and circumvents all the problems related to antibody immobilisation (loss of affinity and correct orientation of the antibody) and is also used when enzyme conjugated primary antibodies are not available for the selected analyte. In any case, the general strategy for immunosensor construction is to place the biological material in close contact with the transducer in order to obtain high sensitivity and to minimise the time of measurement. Many more attempts have been carried out to develop sensitive immunochemical methods for BPA.

The enzyme-linked immunosorbent assay (ELISAs) has been developed for estrogenic bisphenols and have proved to be sensitive. Bisphenol compounds are small molecule which can elicit an immune response only when attached to a large carrier such as a protein. Bisphenol compounds need to be conjugated with a carrier protein commonly such as bovine serum albumin. 4,4-Bis(4-hydroxyphenyl) valeric acid is commercially available reagent to make antigen. However, many more antibodies against protein are manufactured than bisphenolic compound, because carrier protein is bigger than BPA. Therefore, the interferences of the antibodies against carrier protein must be considered in the absorbance for BPA quantification using ELISA. The monoclonal antibodies and developed a direct ELISA with a limit of detection of 5 μg/L have been produced (Goda et al. 2000). The working range of the assay is 5–500 μg/L. Only two different bisphenolic compounds, those who have either one or two hydrogen atoms instead of the methyl groups, are well recognized, whereas metabolites of BPA and other related compounds do not interact with the antibodies. These high cross-reactivities are not a problem, since both compounds are produced in much lower amounts than BPA. An optimized ELISA kit with monoclonal antibodies has also been commercialized by Takeda Chemical Industries. In this case, a more sensitive assay has been achieved with a dynamic range of 0.05–10 μg/L.

Several electrochemical techniques (potentiometric, amperometric, and conductimetric) can be applied for analytical purposes; however, amperometric detection systems have been demonstrated to be the most suitable means for immunosensor construction due to their high sensitivity, low cost and the possibility of instrument miniaturisation. Amperometric detection is based on the measurement of a current at a fixed (potentiostatic technique) or variable (voltammetric technique) potential and classically involves a three-electrode system, although this is often reduced in practice to two electrodes in many devices. By applying a certain potential between the working and the reference electrode, the species of interest is either oxidised or reduced at the working electrode causing a transfer of electrons which ultimately results in a measurable current that is directly proportional to the concentration of the electroactive species at the electrode surface over a wide dynamic range. Electrochemical biosensors use electrical signals as output data. Thus, detection of an analyte is related with the changes of electrical signals, for example, the intensity of the current, potential energy and electrical conductivity of the electrode change. Electrochemical biosensors based on tyrosinase have attracted many interests for the advantages of good reliability, fast response, inexpensive instrument, low energy consumption, simple operation, time saving and high sensitivity. In the presence of dissolved oxygen, tyrosinase could catalyze the oxidation of phenolic compounds to o-diphenols. Biosensor based on the oxidation signal of BPA was fabricated by immobilizing tyrosinase on multiwalled carbon nanotubes-cobalt phthalocyanine-silk fibroin composite modified glassy carbon electrode. In this composite film, silk fibroin provided a biocompatible microenvironment for the tyrosinase to retain its bioactivity, nanotubes possessed excellent inherent conductivity to enhance the electron transfer rate and cobalt phthalocyanine showed good activity to electrooxidation of BPA. The oxidation current was proportional to BPA concentration in the range from 5×10^{-8} to 3×10^{-6} M with correlation coefficient of 0.9979 and detection limit of 3×10^{-8} M (Yin et al. 2010).

To detect BPA in environmental samples, aptamers can be used that are single-stranded oligonucleotides that bind to targets with high affinity and selectivity. Their use as molecular recognition elements has emerged as a viable approach for biosensing. Aptamers offer advantages over other synthetically created molecular recognition systems such as molecular imprinted polymers. For BPA detection, several aptasensing strategies have been developed based on colorimetry, electrochemistry (Zhu et al. 2015), surface-enhanced Raman spectroscopy (Marks et al. 2014) and fluorescence method (Duan et al. 2015). Among them, electrochemical aptamer-based method has attracted great attention because of the favourable attributes of electrochemical methods (low cost, simplicity, rapidity, high sensitivity, and low power requirement). The detection of BPA is mainly based on the competitive recognition of BPA by the immobilized aptamer on the surface of the electrode. The electrochemical aptasensor enables BPA to be detected in drinking water with a limit of detection as low as 0.284 pg/mL in less than 30 min (Xue et al. 2013). The electrochemical aptamer-based sensors for BPA detection were based on one signal, either the type of "signal-on". Based on a triple-signalling strategy, a novel electrochemical aptasensor has been developed for sensitive and selective detection of BPA (Yu et al. 2016). The thiolated ferrocene (Fc)-modified BPA-binding aptamer probe (Fc-) was immobilized on the gold electrode and then hybridized with the methylene blue(MB)-modified complementary DNA probe (MB-P) to form a rigid double-stranded DNA (ds-DNA). The interaction between BPA and Fc-P led to the release of MB-P from the sensing interface and the conformational change of Fc-P. By superimposing the triple signal changes, BPA was detected sensitively with a linear range from 1 to 100 pM. The detection limit is 0.19 pM, and much lower than that obtained by most of the reported electrochemical methods. By changing the specific aptamers, this strategy could be easily extended to detect other redox targets.

References

Ahmadi F, Assadi Y, Milani Hosseini SMR et al (2006) Determination of organophosphorus pesticides in water samples by single drop microextraction and gas chromatography-flame photometric detector. J Chromatogr A 1101:307–312

Arnot JA, Gobas FAPC (2006) A review of bioconcentration factor (BCF) and bioaccumulation factor (BAF) assessments for organic chemicals in aquatic organisms. Environ Rev 14:257–297

Arthur CL, Pawliszyn J (1990) Solid-phase microextraction with thermal desorption using silica optical fibers. Anal Chem 62:2145–2148

Barnes KK, Kolpin DW, Focazio MJ et al (2008a) Water-quality data for pharmaceuticals and other organic wastewater contaminants in ground water and in untreated drinking water sources in the United States, 2000-01. U.S. Geological Survey and U.S. Department of Interior

Barnes KK, Kolpin DW, Furlong ET et al (2008b) A national reconnaissance of pharmaceuticals and other organic wastewater contaminants in the United States—I) groundwater. Sci Total Environ 402:192–200

Boese BL, Winsor M, Lee H II et al (1995) PCB congeners and hexachlorobenzene biota sediment accumulation factors for *Macoma nasuta* exposed to sediments with different total organic carbon contents. Environ Toxicol Chem 14:303–310

Boyd GR, Reemtsma H, Grimm DA et al (2003) Pharmaceuticals and personal care products (PPCPs) in surface and treated waters of Louisiana, USA and Ontario, Canada. Sci Total Environ 311:135–149

Brix R, Postigo C, Gonzalez S et al (2010) Analysis and occurrence of alkylphenolic compounds and estrogens in a European river basin and an evaluation of their importance as priority pollutants. Anal Bioanal Chem 396:1301–1309

Clara M, Kreuzinger N, Strenn B et al (2005) The solids retention time a suitable design parameter to evaluate the capacity of wastewater treatment plants to remove micropollutants. Water Res 39:97–106

Commission Directive 2011/8/EU of 28 January 2011 amending Directive 2002/72/EC as regards the restriction of use of bisphenol A in plastic infant feeding bottles

Covaci A, Voorspoels S, Abdallah MAE et al (2009) Analytical and environmental aspects of the flame retardant tetrabromobisphenol-A and its derivatives. J Chromatogr A 1216:346–363

Crain DA, Eriksen M, Iguchi T et al (2007) An ecological assessment of bisphenol-A: evidence from comparative biology. Reprod Toxicol 24:225–239

Cunha SC, Fernandes JO (2010) Quantification of free and total bisphenol A and bisphenol B in human urine bydispersive liquid–liquid microextraction (DLLME) and heart-cutting multidimensional gas chromatography–mass spectrometry (MD–GC/MS). Talanta 83:117–125

Directive 2003/89/EC of the European Parliament and of the Council of 10 November 2003 amending Directive 2000/13/EC as regards indication of the ingredients present in foodstuffs

Dorn PB, Chou CS, Gentempo JJ (1987) Degradation of bisphenol A in natural waters. Chemosphere 16:1501–1507

Duan N, Zhang H, Nie Y et al (2015) Fluorescence resonance energy transfer-based aptamer biosensors for bisphenol A using lanthanide-doped KGdF 4 nanoparticles. Anal Methods 7:5186–5192

EFSA (European Food Safety Authority) (2010) Scientific opinion on bisphenol A: evaluation of a study investigating its neurodevelopmental toxicity, review of recent scientific literature on its toxicity and advice on the Danish risk assessment of bisphenol A. EFSA J 8:1829. Available from: URL:http://www.efsa.europa.eu/en/efsajournal/pub/1829.htm. Accessed at 8 May 2014

Eisenreich SJ, Looney BB, Thornton JD (1981) Airborne organic contaminants in the Great Lakes ecosystem. Environ Sci Technol 15:30–38

Environmental Working Group (2010) BPA in store receipts. Available from http://www.ewg.org/node/28589/print

European Parliament and The Council (2006) Directive 2006/11/EC on pollution caused by certain dangerous substances discharged into the aquatic environment of the community, 15 February 2006

European Union (2010) Updated European Risk Assessment Report 4,4-Isopropylidenediphenol (Bisphenol-A). Commission Directive 2011/8/EU, 2011 of 28 January 2011 amending Directive 2002/72/EC as regards the restriction of use of bisphenol A in plastic infant feeding bottles

Fent G, Hein WJ, Moendel MJ et al (2003) Fate of 14C-bisphenol A in soils. Chemosphere 51:735–746

Flint S, Markle T, Thompson S et al (2012) Bisphenol A exposure, effects, and policy: a wildlife perspective. J Environ Manag 104:19–34

Focazio MJ, Kolpin DW, Barnes KK et al (2008) A national reconnaissance for pharmaceuticals and other organic wastewater contaminants in the United States—II) Untreated drinking water sources. Sci Total Environ 402:201–216

Funakoshi G, Kasuya S (2009) Influence of an estuary dam on the dynamics of bisphenol A and alkylphenols. Chemosphere 75:491–497

Fung EYK, Ewoldsen NO, St Germain HA et al (2000) Pharmacokinetics of bisphenol A released from a rental sealant. J Am Dent Assoc 131:51–58

Fürhacker M, Scharf S, Weber H (2000) Bisphenol A: emissions from point sources. Chemosphere 41:751–756

Geens T, Goeyens L, Covaci A (2011) Are potential sources for human exposure to bisphenol-A overlooked? Int J Hyg Environ Health 214:339–347

Ghosh NN, Kiskan B, Yagci Y (2007) Polybenzoxazines—new high performance thermosetting resins: synthesis and properties. Prog Polym Sci 32:1344–1391

Goda Y, Kobayashi A, Fukuda K et al (2000) Development of the ELISAs for detection of hormone-disrupting chemicals. Water Sci Technol 42:81–88

Golub MS, Wu KL, Kaufman FL et al (2010) Bisphenol A: developmental toxicity from early prenatal exposure. Birth Defects Res B 89:441–466

Hanze (1994) Bisfenol A—ecotoxikologisk faroanalys. Vetenskaplig Utredningoc Dokumentation. Kem I. Kemikalienspektionen. In: TemaNord. 1996

Hohenblum P, Gans O, Moche W et al (2004) Monitoring of selected estrogenic hormones and industrial chemicals in ground waters and surface waters in Austria. Sci Total Environ 333:185–193

HSDB (2000) Hazardous substances database

Jackson J, Sutton R (2008) Sources of endocrine-disrupting chemicals in urban wastewater, Oakland, CA. Sci Total Environ 405:153–160

Kinney CA, Furlong ET, Zaugg SD et al (2006) Survey of organic wastewater contaminants in biosolids destined for land application. Environ Sci Technol 40:7207–7215

Kinney CA, Furlong ET, Kolpin DW et al (2008) Bioaccumulation of pharmaceuticals and other anthropogenic waste indicators in earthworms from agricultural soil amended with biosolid or swine manure. Environ Sci Technol 42:1863–1870

Klečka GM, Staples CA, Clark KE et al (2009) Exposure analysis of bisphenol A in surface water systems in North America and Europe. Environ Sci Technol 43:6145–6150

Kolpin DW, Furlong ET, Meyer MT et al (2002) Pharmaceuticals, hormones, and other organic wastewater contaminants in US streams, 1999–2000: a national reconnaissance. Environ Sci Technol 36:1202–1211

Kuch HM, Ballschmiter K (2001) Determination of endocrine-disrupting phenolic compounds and estrogens in surface and drinking water by HRGC-(NCI)-MS in the picogram per liter range. Environ Sci Technol 35:3201–3206

Lee HB, Peart TE (2000a) Bisphenol A contamination in Canadian municipal and industrial wastewater and sludge samples. Water Qual Red J Can 35:283–298

Lee HB, Peart TE (2000b) Determination of Bisphenol A in sewage effluent and sludge by solid-phase and supercritical fluid extraction and gas chromatography/mass spectrometry. J AOAC Int 83:290–297

Lee CC, Jiang LY, Kuo YL et al (2013) The potential role of water quality parameters on occurrence of nonylphenol and bisphenol A and identification of their discharge sources in the river ecosystems. Chemosphere 91:904–911

Lee CC, Jiang LY, Kuo YL et al (2015) Characteristics of nonylphenol and bisphenol A accumulation by fish and implications for ecological and human health. Sci Total Environ 502:417–425

Li Y, Gaoa Y, Cao Y et al (2012) Electrochemical sensor for bisphenol A determination based on MWCNT/melamine complex modified GCE. Sens Actuators B 171–172:726–733

Li Y, Zhai X, Liu X et al (2016) Electrochemical determination of bisphenol A at ordered mesoporous carbon modified nano-carbonionic liquid paste electrode. Talanta 148:362–369

Mackintosh CE, Maldonado J, Hongwu J et al (2004) Distribution of phthalate esters in a marine aquatic food web: comparison to polychlorinated biphenyls. Environ Sci Technol 38:2011–2020

Marks HL, Pishko MV, Jackson GW (2014) Rational design of a bisphenol A aptamer selective surface-enhanced Raman scattering nanoprobe. Anal Chem 86:11614–11619

Melcer H, Klečka G (2011) Treatment of wastewaters containing bisphenol A: state of the science review. Water Environ Res 83:650–666

Nordic Council of Ministers (1996) Chemical with estrogen-like effects. TemaNord 580

O'Neil MJ (ed) (2006) The Merck index: an encyclopedia of chemicals, drugs, and biologicals (14th ed). NJ: Merck, 211, as cited in the Hazardous Substances Data Bank (HSDB), a database of the National Library of Medicine's TOXNET system, Bisphenol A

Osenbrück K, Gläser HR, Knöller K et al (2007) Sources and transport of selected organic micropollutants in urban ground water underlying the city of Halle (Saale), Germany. Water Res 41:3259–3270

Pedersen-Bjergaard S, Rasmussen KE (1999) Liquid-liquid-liquid microextraction for sample preparation of biological fluids prior to capillary electrophoresis. Anal Chem 71:2650–2656

Peltonen K, Pfäffli P, Itkonen A et al (1986) Determination of the presence of bisphenol-A and the absence of diglycidyl ether of bisphenol-A in the thermal degradation products of epoxy powder paint. Am Ind Hyg Assoc J 47:399–403

Perez RM, Sandler JKW, Altstadt V et al (2007) Novel phosphorus modified polysulfone as a combined flame retardant and toughness modifier for epoxy resins. Polymer 48:778–790

Reinstorf F, Strauch G, Schirmer K et al (2008) Mass fluxes and spatial trends of xenobiotics in the waters of the city of Halle, Germany. Environ Poll 152:452–460

Rezaee M, Assadi Y, Milani Hosseini MR et al (2006) Determination of organic compounds in water using dispersive liquid–liquid microextraction. J Chromatogr A 1116:1–9

Rubin BS (2011) Bisphenol A: an endocrine disruptor with widespread exposure and multiple effects. J Steroid Biochem Mol Biol 127:27–34

Rudel RA, Melly SJ, Geno PW et al (1998) Identification of alkylphenols and other estrogenic phenolic compounds in wastewater, septage, and groundwater on Cape Cod, Massachusetts. Environ Sci Technol 32:861–869

Sakai K, Yamanaka H, Moriyoshi K et al (2007) Biodegradation of bisphenol A and related compounds by Sphingomonas sp strain BP-7 isolated from seawater. Biosci Biotechnol Biochem 71:51–57

Satoh K, Ohyama K, Aoki N et al (2004) Study on anti-androgenic effects of bisphenol a diglycidyl ether (BADGE), bisphenol F diglycidyl ether (BFDGE) and their derivatives using cells stably transfected with human androgen receptor, AR-EcoScreen. Food Chem Toxicol 42:983–993

Staples CA, Dorn PB, Klecka GM et al (1998) A review of the environmental fate, effects, and exposures of bisphenol A. Chemosphere 36:2149–2173

Staples C, Friederich U, Hall T et al (2010) Estimating potential risks to terrestrial invertebrates and plants exposed to bisphenol A in soil amended with activated sludge biosolids. Environ Toxicol Chem 29:467–475

Takeuchi I, Miyoshi N, Mizukawa K et al (2009) Biomagnification profiles of polycyclic aromatic hydrocarbons, alkylphenols and polychlorinated biphenyls in Tokyo Bay elucidated by d13C and d15N isotope ratios as guides to trophic web structure. Mar Pollut Bull 58:663–671

Tan F, Zhao H, Li X et al (2009) Preparation and evaluation of molecularly imprinted solid-phase microextraction fibers for selective extraction of bisphenol A in complex samples. J Chromatogr A 1216:5647–5654

Tomy GT, Budakowski W, Halldorson T et al (2004) Biomagnification of α- and γ-hexabromocyclododecane isomers in a Lake Ontario food web. Environ Sci Technol 38:2298–2303

Tsai WT (2006) Human health risk on environmental exposure to bisphenol-A: A review. J Environ Sci Health C 24:225–255

Thevenot DR, Toth K, Durstand RA et al (1999) Electrochemical biosensors: recommendation definition and classification. Pure Appl Chem 71:2333–2348

Thevenot DR, Toth K, Durst RA et al (2001) Biosens Bioelectron 16:121–131

US EPA (1989) Risk assessment guidance for superfund. Volume I, human health evaluation manual—part A. EPA/540/1-89/002, December, 1989. Office of Emergency and Remedial Response, Washington, DC, p 20450. Available from: http://www.epa.gov/oswer/riskassessment/ragsa/index.htm

US EPA (1997) Exposure factors handbook. Washington DC, EPA/600/P-95/002F a-c

US EPA (1998a) Guidelines for ecological risk assessment. Risk assessment forum. Washington DC [EPA/630/R095//002 F]

US EPA (1998b) Coatings and consumer products group emission standards division office of air quality planning and standards research triangle park, NC 27711

US EPA (2008) Report on the environment (ROE)

US EPA (2010) Bisphenol A Action Plan (CASRN 80-05-7) [CA Index Name: Phenol, 4,4′-(1-methylethylidene)bis, 3/29/2010

US EPA (2011) Advanced notice of proposed rulemaking: testing of bisphenol A. Fed Reg 76 (143):44535–44547

Van der Oost R, Beyer J, Vermeulen NPE (2003) Fish bioaccumulation and biomarkers in environmental risk assessment: a review. Environ Toxicol Pharmacol 13:57–149

Vandenberg LN, Hauser R, Marcus M et al (2007) Human exposure to bisphenol A (BPA). Reprod Toxicol 24:139–177

Vogel SA (2009) The politics of plastics: the making and unmaking of bisphenol A "safety". Am J Public Health 99:S559–S566

Wang YQ, Zhang HM (2015) Exploration of binding of bisphenol A and its analogues with calf thymus DNA by optical spectroscopic and molecular docking methods. J Photochem Photobiol B: Biology 149:9–20

Wang F, Yang J, Wu K (2009) Mesoporous silica-based electrochemical sensor for sensitive determination of environmental hormone bisphenol A. Anal Chim Acta 638:23–28

Wang YQ, Zhang HM, Cao J (2014) Quest for the binding mode of tetrabromobisphenol A with Calf thymus DNA. Spectrochim Acta A 131:109–113

Watabe Y, Kondo T, Morita M et al (2004) Determination of bisphenol A in environmental water at ultra-low level by high-performance liquid chromatography with an effective on-line pretreatment device. J Chromatogr A 1032:45–49

Welshons WV, Nagel SC, vom Saal FS (2006) Large effects from small exposures. III. Endocrine mechanisms mediating effects of bisphenol A at levels of human exposure. Endocrinology 147: S56–S69

Wenzel A, Müller J, Ternes T (2003) Study on endocrine disrupters in drinking water final report ENV.D.1/ETU/2000/0083, 26 February

Wetherill Y, Akingbemi B, Kanno J et al (2007) In vitro molecular mechanisms of bisphenol A action. Reprod Toxicol 24:178–198

Wilson NK, Chuang JC, Morgan MK et al (2007) An observation study of the potential exposures of preschool children to pentachlorophenol, bisphenol A, and nonylphenol at home and daycare. Environ Res 103:9–20

Wong CS, Capel PD, Nowell LH (2001) National-scale, field-based evaluation of the biota–sediment accumulation factor model. Environ Sci Technol 35:1709–1715

Wright-Walters M, Volz C, Talbott E et al (2011) An updated weight of evidence approach to the aquatic hazard assessment of bisphenol A and the derivation a new predicted no effect concentration (Pnec) using a non-parametric methodology. Sci Total Environ 409:676–685

Xue F, Wu JJ, Chu HQ et al (2013) Electrochemical aptasensor for the determination of bisphenol A in drinking water. Microchim Acta 180:109–115

Yin H, Zhou Y, Xu J et al (2010) Amperometric biosensor based on tyrosinase immobilized onto multiwalled carbon nanotubes-cobalt phthalocyanine-silk fibroin film and its application to determine bisphenol A. Anal Chim Acta 659:144–150

Ying GG, Kookana RS, Dillon P (2003) Sorption and degradation of selected five endocrine disrupting chemicals in aquifer material. Water Res 37:3785–3791

Yu P, LiuY Zhang X et al (2016) A novel electrochemical aptsensor for bisphenol A assay based on triple-signalling strategy. Biosens Bioelectron 79:22–28

Zhao JL, Ying GG, Wang L et al (2009) Determination of phenolic endocrine disrupting chemicals and acidic pharmaceuticals in surface water of the Pearl Rivers in South China by gas chromatography-negative chemical ionization-mass spectrometry. Sci Total Environ 407:962–974

Zhu Y, Zhou C, Yan X et al (2015) Aptamer-functionalized nanoporous gold film for high-performance direct electrochemical detection of bisphenol A in human serum. Anal Chim Acta 883:81–89

Chapter 3
Physical and Chemical Treatment Technologies for BPA Removal from Wastewater

In recent years, there have been intensive efforts toward the development of adsorption for the removal of phenolic endocrine disrupting chemicals, such as BPA, from aqueous matrices. Advanced oxidation processes are also likely to play a key role; recent attempts have dealt with BPA degradation by ozonation, ultrasound irradiation, dark- and photo-Fenton oxidation, and electrochemical oxidation. In addition, the use of membrane filtration for BPA removal from water and wastewater has been reported. Although these physicochemical methods are effective in the removal of endocrine disrupting compounds (EDCs), their low cost-effectiveness in dealing with large volume of low-level pollutants and the risk of generation of toxic by-products constitute major barriers in the field application.

3.1 Adsorption

In wastewater treatment and water reclamation, adsorption is integral to a broad spectrum of physical, chemical, and biological processes and operations. Adsorption takes place via a variety of mechanisms. These mechanisms include the forces active within the boundaries of phases or surfaces; these boundary forces not only cause adsorption, but are influenced by adsorption, creating the potential for feedback effects. These forces result in characteristic boundary energies. The extent of adsorption is generally proportional to a specific surface area. The adsorbents can take a broad range of chemical forms and different geometrical surface structures. This is reflected in the range of their applications.

BPA is removed by adsorption on activated carbons, carbon nanomaterials, zeolites and minerals, hybrid particles, inorganic-organic modified bentonite and mesoporous silica-based materials. While these adsorbents have high adsorption capacities, they suffer from some inherent limitations, such as the separation of very small-sized materials from the treated matrix. Recently, a great attention has been focused on the application of nano-structured materials as adsorbents of organic substances from wastewater (Aditya et al. 2011; Mohmood et al. 2013).

M. ZIELIŃSKA et al., *Bisphenol A Removal from Water and Wastewater*,
https://doi.org/10.1007/978-3-319-92361-1_3

Activated carbon (AC) is a commonly used material for adsorption. The high specific area and low surface polarity of AC is the main factor resulting in a high adsorption of BPA (Tsai et al. 2006b). Major disadvantages of using the AC are the disordered structure, limitation of pore size to micropores (<2 nm) and irregular pore size distribution. The micropores could limit the mass transfer or decrease pore accessibility for larger adsorbates. The adsorption of BPA depends fundamentally on the surface nature of AC and the chemical properties of the solution (Bautista-Toledo et al. 2005). The most favourable conditions are those in which the net charge density of the carbon is zero and the BPA is in molecular form. The presence of mineral matter in carbons reduces adsorption capacity because of the hydrophilic nature of the matter. The presence of electrolytes in the solution favours adsorption because they cause the surface of the carbon to become positively charged, which increases adsorbent-adsorbate interactions. Thus, although the net charge of the carbon should be zero, a positive surface charge is desirable.

The capacity of AC to adsorb BPA can be improved by increasing its content of acidic oxygen-containing groups and its surface charge density via chemical modification, oxidation or thermal treatment. Liu et al. (2009) used commercial ACs that had been selectively modified with nitric acid and thermal treatment. The highest equilibrium adsorption amounts of BPA reached 382.12 and 432.34 mg/g, respectively for non-treated and thermally modified carbons. BPA adsorption was found to be temperature- and pH-dependent: the adsorbed amount of BPA decreased with the increase of temperature from 288 to 318 K; at pH 11.0 ACs represented the weakest adsorption capacity.

There are relationships between the carbonization temperature of woody carbon, such as bamboo charcoal and Sugi charcoal, and the removal of BPA. The adsorption properties of chemicals are different, based on the carbonization temperature. The adsorption constant *K* by Freundlich equation for the produced carbonaceous materials at a carbonization temperature of 1073 K is greater than that at 873 K (Nakanishi et al. 2002). The constant *K* was also larger in the order of the AC; Sugi sawdust, Sugi chips, and Hinoki sawdust. The more closely the BPA molecules in the pores are located to the surrounding pore walls, the higher will be the adsorption force. The activation temperature increases the iodine adsorption capacity and the adsorption force. The carbonaceous materials, which are produced from organic by-products, could be used as adsorbents.

Porous carbon materials are important adsorbents. The popularity of these materials can be attributed to their high specific area, large pore volume, chemical inertness and good mechanical stability. The first hard templated ordered mesoporous that were synthesized, used MCM-48 mesoporous silica molecular sieves as template. The resulting replicated mesoporous carbon CMK-1 exhibited porous structures consisting of two disconnected interwoven three-dimensional pore systems (Kruk et al. 2000). Soft templated synthesis of ordered mesoporous materials by self-assembly of tri-block copolymer surfactants and carbon precursor was pioneered by Dai et al. (1996). Mesopores of soft templated carbon and hard templated carbon enhance the BPA diffusion into the adsorbent (Libbrecht et al. 2015).

The hard templated carbon (CMK-3) material had higher BET-surface of 1420 m^2/g with an average pore size of 4 nm. The soft templated carbon (SMC) reached a BET-surface of 476 m^2/g and a pore size of 7 nm. The maximum observed adsorption capacity of CMK-3 was the highest with 474 mg/g, compared to 290 mg/g for activated carbon (PAC) and 154 mg/g for SMC. The difference in adsorption capacities was attributed to the specific surface area and hydrophobicity of the adsorbent. The difference in adsorption kinetics is caused by the increase in pore diameter. Hard templated carbon with an open geometry consisting of interlinked nanorods allows for even faster intraparticle diffusion (Libbrecht et al. 2015).

The carbon nanomaterials can serve as adsorbents because of their hollow and layered structure and large specific surface area, which is why they are the most commonly used nano-materials. In contrast to AC, carbon-based nano-sized sorbents such as carbon nanotubes or graphene may offer a more physically homogeneous surface with a high surface area 150–3000 m^2/g, shorter equilibrium times without pore diffusion restrictions, all of which could lead to a more rapid or greater extent of sorption for organic compounds. Compared to conventional AC, graphene offers higher sorption capacities and faster equilibration over a wide pH range, due to more homogeneous sites, less pore diffusion, and rapid π–π interaction kinetics. Graphene's high specific surface area, theoretically 2630 m^2/g (Zhu et al. 2011), but in practice often <1000 m^2/g, may promote graphene-based materials as a carbonaceous sorbent for interfacial BPA contaminant reactions. Parameters such as the pore diameter distribution and adsorption energy distribution are needed to quantify adsorption on AC, while, for carbon nanomaterials, one can deal directly with the various well-defined adsorption sites available to the adsorbed molecules.

Carbon nanotubes (CNTs) provide a large specific surface area and a strong van der Waals binding energy for molecular adsorbates on well-defined adsorption sites such as interior sites, groove sites, exterior sites and interstitial sites (Zhang et al. 2009). The binding energy of the interior sites is the highest, followed by the one of the groove sites on the outside of carbon nanotubes bundles, and the one of the exterior sites on the convex outer surface is the lowest (Ren et al. 2011). The interstitial sites appear to be inaccessible to the adsorbate molecules. From the viewpoint of CNT structure, the relationship between them and other carbonaceous adsorptive materials is almost the same as that between single crystals and polycrystalline materials (Kondratyuk and Yates 2007). The use of single-walled CNTs as an adsorbent for pretreatment of seawater and brackish water, prior to desalination and removal of BPA was evaluated (Joseph et al. 2011). The adsorptive capacity of these nanotubes was low for BPA. The decrease in adsorption of BPA may be a result of the increased ionization caused by the increase in pH, which can lead to reduced hydrophobic interactions with the single-walled carbon nanotubes. If electron donor-acceptor π–π interactions contribute to the adsorption of BPA, decreased adsorption can be expected when the pH is higher than the dissociation constant (*pKa*), because the chemical and the nanotubes are both negatively charged and experience increased electrostatic repulsion (Pan and Xing 2008). The surface charge (expressed as zeta potential, mV) of the single-walled CNTs was negative. The adsorption of BPA remains fairly constant until the pH exceeds *pKa* for BPA, at which point the adsorption is reduced. Changes in the water chemistry did not significantly affect the overall adsorption, while changing the water pH from 3.5 to 11 showed a reduction in the adsorption of BPA. As the concentration of natural organic matter (NOM) increases in water sources, overall adsorption of BPA is expected to be increasingly affected, possibly due to direct competition for adsorption sites on the single-walled CNTs or pore blockage by the NOM. Significant removal of dissolved organic carbon (DOC) with increasing doses of nanotubes due to similar mechanisms that govern the adsorption of BPA (e.g. hydrophobic and π–π interactions) was observed by Joseph et al. (2011). The adsorption affinity of pollutants depends on not only the surface area of adsorbents but also the pore size distribution of adsorbents. Based on the studies of adsorption isotherms and pH effect, multiple factors such as π–π interaction, hydrophobic interaction and electrostatic interaction played important roles in the adsorption of ciprofloxacin, BPA and 2-chlorophenol on electrospun carbon nanofibers (Li et al. 2015a). Adsorption capacities of these three pollutants on electrospun carbon nanofibers were in the trend of 2-chlorophenol > BPA > ciprofloxacin, and increased 1.1–2.6-fold, compared with that of commercial available AC.

The small size (nanoscale particles) of CNTs makes them difficult to separate from aqueous solution. Ultracentrifugation and membrane filtration are efficient techniques to separate CNTs from aqueous solutions (Ren et al. 2011). Compared with centrifugation and filtration methods, the magnetic separation method is considered a rapid and effective technique for separating nanoparticles from aqueous solution. The CNTs/magnetic composites are promising materials in environmental pollution management in large scale. Functional magnetic nanomaterials are synthesized by anchoring immobilized polymer, inorganic or organic molecules. Magnetic nanoadsorbents can bind with non-magnetic target molecules via some intermediates forming complex which can be easily separated and recovered from multiphase systems, opening the door for sorbent regeneration, safe disposal of waste and/or recovery (Youn et al. 2009). More attentions have directed to the combination of magnetic materials and CNTs (Kim et al. 2010). But the cost of CNTs was high and CNTs are difficult to be obtained. The substitute of CNTs should arouse people's attention the supporter of molecularly imprinted polymers (MMIP). Molecular imprinting technique is an efficient method of producing a 3D cross-linked polymer network in the material that preferentially binds with a specific template. The supporter of molecularly imprinted polymers (MMIPs) was prepared by using kaolinite/Fe_3O_4 (KLT/Fe_3O_4). Kaolinite is a kind of clay, which has many advantages such as wide sources, low cost, simple process and easy use. The behaviour of the MMIPs for the separation and recognition of BPA from aqueous solution was tested (Guo et al. 2011). The BPA removal by the MMIPs was a combination of selective adsorption and magnetic separation. The adsorption capacities of BPA for the second and the third step maintained more than 85% of that for the first adsorption step, which illustrated that the MMIPs have the potentiality to be reused Guo et al. (2011). BPA removal from aqueous solutions was examined by batch adsorption onto 6-deoxy-6-ethylenediamino-b-cyclodextrin (b-CDen) grafted thiodiglycolic acid (TDGA) modified magnetic nanoparticles

(CDen-MNPs) (Ghosh et al. 2013). b-CDen being grafted on TDGA-coated Fe_3O_4 nanoparticle contributed to an enhancement of adsorption capacities because of the inclusion abilities of its hydrophobic cavity with organic contaminants through host–guest interactions. Physicochemical properties as *pKa* and hydrophobicity were important factors in adsorption process. BPA (octanol/water partition coefficient log $K_{ow} = 3.3$) was least adsorbed, whereas carbamazepine (log $K_{ow} = 2.45$) was most adsorbable. Van der Waal's interaction was found to be dominant.

Natural polymer chitosan, produced commercially by deacetylation of chitin, is a linear polysaccharide composed of randomly distributed β-(1-4)-linked D-glucosamine and *N*-acetyl-D-glucosamine, which can be used for BPA removal from water. Chitosan was crosslinked with glutaraldehyde and subsequently treated with α-ketoglutaric acid to form Schiff's base. Schiff's base was reduced to produce α-ketoglutaric acid-modified chitosan resins. With these resins as weak cation exchanger, a new hydrophobic sorption mechanism of hemimicelles for sorbing BPA from aqueous solution was investigated. The sorption capacity of modified resins for BPA was 30.67 mg/g. Novel hydrogel beads having molecular adsorption abilities were prepared from carboxymethylcellulose sodium salt (CMC) and β-cyclodextrin (β-CD) by suspension crosslinking, using ethylene glycol diglycidyl ether (EGDE) in basic medium as a crosslinking agent (Kono et al. 2013). The cyclodextrins (CDs) are easily complexed with many chemical species, including organic molecules, inorganic molecules, and within their cavities. The CDs are torus-shaped cyclic oligosaccharides composed of α-(1 → 4) linked d-glucopyranose with six, seven, and eight units, namely, α-CD, β-CD, and γ-CD, respectively. The binding mechanisms are attributed to chemical bonds or weak intermolecular forces. The hydroxyl groups of CDs can undergo reactions such as etherification, esterification, reduction, and crosslinking to obtain derivatives with properties, such as high water solubility, ideal surface activity, and lower water absorption. The hydrogel beads (CDs) showed a high adsorption capacity toward BPA in water. The maximum BPA-adsorption by hydrogel beads amounted to 167 μmol/g (Kono et al. 2013).

A significant aspect of the inorganic adsorbents is that the bonding forces between the adsorbent and the adsorbate are usually weaker than those encountered in activated carbon or polymer adsorbents. The regeneration of the inorganic adsorbents can be accomplished by simple and non-destructive methods, such as solvent washing or calcining, which provides the potential approaches to recover these adsorbents. Zeolite synthesized from coal fly ash was modified with hexadecyltrimethylammonium (HDTMA) and was examined for the BPA adsorption from water. HDTMA was confirmed to form bilayer micelles on external surfaces of zeolites. Uptake of BPA was greatly influenced by pH; alkaline conditions enabled the deprotonation of BPA to form organic anions. BPA anions interact strongly with the positively charged heads of HDTMA, with the two hydrophobic benzene rings of BPA pointing to the inside of HDTMA bilayers. The adsorption of uncharged BPA probably involved hydrophobic partitioning into HDTMA bilayers and the coordination of the oxygen atoms of BPA with positively charged heads of HDTMA (Dong et al. 2010b). Hydrophobic zeolite inorganic adsorbent can be characterized as absorbent with significant surface and pore characteristics because of its high ratio of silica to alumina (i.e. Si/Al or SiO_2/Al_2O_3) and low hydroxyl content. The dealuminated zeolite has been studied as an effective adsorbent for removal of organic compounds from an aqueous solution. The adsorption behaviors depend greatly on the initial BPA concentration and adsorbent dosage. The effect of solution pH on the equilibrium adsorption capacity, however, is only significant at very basic pH values, probably due to the deprotonation of BPA and resulting electrostatic interactions (Tsai et al. 2006a). The organic Arizona SAz-2 Ca-montmorillonite was directly intercalated with two long chain cationic surfactants (DDTMA and HDTMA) without prior exchange with Na^+. These materials possess excellent adsorption capacity for BPA from aqueous solution (Zheng et al. 2013). The hydrophobic phase and positively charged surface created by the loaded surfactant molecules are responsible for the adsorption of BPA. The prepared organoclays intercalated with more surfactants or longer chain surfactant molecules were less influenced by the solution pH because of the increase of positive surface charge. The adsorption reaction of BPA by organoclays was a spontaneous exothermic process. The organic Arizona Ca-MMT modified by long chain surfactants is a promising adsorbent for environmental pollutants.

3.2 Advanced Oxidation Processes (AOP)

Advanced oxidation processes (AOPs) are based on the generation of hydroxyl radicals ($^{\cdot}OH$) that interact with the pollutants and degrade them into by-products with lower molecular weight and they can even achieve their complete mineralization. For this reason, AOPs are the effective techniques for the degradation of organic compounds such as BPA, in aqueous solution. Several AOPs have been utilized, including Fenton and Fenton-related processes, photolysis, photocatalysis, ozonation, and various combinations of these.

3.2.1 Ozonation

The reaction of ozone with BPA takes place either via direct reaction with molecular ozone or indirect reactions with hydroxyl radicals $^{\cdot}OH$ produced by the decomposition of ozone in alkaline conditions (Rokhina and Virkutyte 2010). This results in the breakdown of high molecular weight compounds to low molecular weight compounds and improvement of the biodegradability of the treated water. The ozone molecule reacts selectively with compounds containing C≡C bonds, certain functional groups (e.g. –OH, $–CH_3$, $–OCH_3$) and anions (those of N, P, O, S); oxidation by $^{\cdot}OH$ is non-selective. Direct reaction of ozone is known to occur under acidic conditions and in the presence of radical scavengers that inhibit the chain reaction responsible for ozone decomposition. Under alkaline conditions that promote the radical-type chain reaction and $^{\cdot}OH$ formation, the indirect reaction predominates due to the extremely rapid and non-selective nature of $^{\cdot}OH$ (10^9 1/(M s)) (Ning et al. 2007). At pH 7 and 20 °C, the apparent second-order rate constant of 2.7×10^6 1/(M s) has been reported (taking into account the difference of second-order reaction rate constant between ozone and molecular BPA and anionic BPA, i.e. 1.68×10^4 1/(M s), 1.06×10^9 1/(M s) and 1.11×10^9 1/(M s), respectively, for BPA, BPA^- and BPA^{2-}. Therefore, for an ozone concentration of 1 mg/L, the half-life of BPA can be evaluated to be lower than 15 ms (Deborde et al. 2005). Due to its electrophilic nature, ozone usually reacts with aromatic rings by electrophilic substitution or 1,3-dipolar cycloaddition (Mvula and von Sonntag 2003), and a similar reaction mechanism is expected to occur during its reaction with BPA (Deborde et al. 2008).

The degradation of BPA by ozone leads to the formation of a range of by-products, as shown by studies in simple water matrices such as MilliQ or deionized water. The formation of major reaction intermediates was studied by Deborde et al. (2008) using BPA (initial concentration 22.8 mg/L) in MilliQ water at pH 6.5. Five major transformation products, namely muconic acid derivatives of BPA, benzoquinone, 2-(4-hydroxyphenyl)-propan-2-ol, orthoquinone and catechol, in addition to several other minor transformation products were identified. The intermediates of catechol, orthoquinone, hydroquinone, and muconic acid derivatives are well known to be the reaction by-products of the ozonation of phenol (Yamamoto et al. 1979). Due to the phenolic ring in BPA, the mechanisms and by-products similar to the reaction of ozone with phenol can be expected (Deborde et al. 2005). Further transformation of these major intermediates would lead to more polar compounds (acids and aldehydes). Similar to ozonation of phenol, progressive formation of by-products with increasing ozone dosage followed by decrease in their formation was observed, which confirmed the low stability of the intermediates and thus demonstrated that different intermediates are dependent on ozone dosage. The minor transformation products are believed to consist of oligomeric structures arising from secondary reaction between various oxidation products of BPA, however, they are more complex to assess. An ozone dosage of 0.57 mM (2.73 mg/L), corresponding to ozone/BPA ratio of 5.7 (slightly higher than needed for complete mineralization of the BPA), was adequate to remove all the major transformation products. It is therefore important to consider the reaction of ozone with the by-products generated in real wastewater situations when there are several other compounds that could react with ozone and ultimately increase the ozone demand (Deborde et al. 2008). For BPA ozonation, the stoichiometric ratio between ozone and BPA was 10.3 at 3.0 pH. The presence of malonic and oxalic acids after ozonation time of 5 min demonstrated that the mineralization of intermediates began simultaneously with the ozonation of BPA. Mineralization of BPA solution was not as successful as degradation of BPA for ozonation times up to 25 min. At the end of 25 min ozonation time, all BPA molecules were completely degraded, while only about 30% of BPA was mineralized. This makes it difficult to compare the efficacy of the treatment in different experiments as no consistent endpoint for BPA removal was used. The efficacy varied from mineralization to a mixture of by-products, the identity of which varied with the conditions (Kusvuran and Yildirim 2013).

The rate of BPA removal from the aqueous solution linearly increased with ozone dosage (Garoma and Matsumoto 2009). The stoichiometry during ozonation of BPA was found to be 2.0–3.4. For complete removal of BPA, the moles of ozone absorbed by the solution per mole of BPA varied in the range of 7.0–9.5. Increasing the pH from 2.0 to 7.0 improved the removal of BPA but an increase to 10.0 resulted in decreased removal efficiency. The situation is further complicated when other organic matter is present in water and wastewater. The effect of various types of organic matter on the removal of BPA by ozone needs to be studied as these organics potentially influence the required ozone dosage, reaction mechanism, toxicity and estrogenicity of the treated water. The mineralization of BPA investigated was using non-catalytic and catalytic ozonation and adsorption in a semi-batch reactor (Keykavoos et al. 2013). Fast disappearance of BPA in the system is in accordance with very high reaction rate constant for reaction between ozone and BPA in the range of 1.3×10^4 1/(M s) and 2.7×10^6 1/(M s). Ozone is able to attack double bounds of aromatic compounds such as BPA but it is unable to react with linear compounds, as BPA by-products. During the reaction between ozone and BPA, pH dropped from 5 to 3.6 after 1 h of ozonation. This can be attributed to the formation of final nondegradable linear acidic by-products during the ozonation process. Long reaction times indicate that non-catalytic ozonation only leads to incomplete mineralization of BPA, producing linear acidic by-products which cannot be removed by ozonation.

Catalytic ozonation can be considered firstly as homogeneous catalytic ozonation, which is based on ozone activation by metal ions present in aqueous solution, and secondly as heterogeneous catalytic ozonation in the presence of metal oxides or metals/metal oxides on supports. Two major mechanisms of homogeneous catalytic ozonation can be noted: decomposition of ozone by metal ions leading to the generation of free radicals (Sauleda and Brillas 2001) and complexes formation between BPA and the catalyst and subsequent oxidation of the complex (Beltran et al. 2005).

Mineralization of BPA refers to the conversion of BPA and its by-products in an aqueous environment to water and carbon dioxide. In order to attain complete BPA mineralization, ozonation process can be performed more efficiently in the presence of an active catalyst. Several metal ions were found to be effective as catalysts in the ozonation process. Among the most widely used are: Mn(II), Fe(III), Fe(II), Co(II), Cu(II), Zn(II) and Cr(III). Fe(II)/O_3 and Fe(II)/O_3/UV processes are based on Fenton or photo-Fenton reactions. Oxidation/reduction reactions play an important role in catalytic ozonation systems utilizing transition metals. Oxidation/reduction reactions characteristic for Fe(II)/O_3 were observed also for Mn(II) ions. Among the most widely used catalysts in heterogeneous catalytic ozonation are metal oxides (MnO_2, TiO_2, Al_2O_3, FeOOH and CeO_2), metals (Cu, Ru, Pt, Co) on supports (SiO_2, Al_2O_3, TiO_2, CeO_2 and activated carbon) and zeolites modified with metals.

Alumina was found to be poor adsorbent of BPA in the absence of ozone in the system, but this catalyst in the presence of ozone can enhance final mineralization of BPA up to 90%. Smaller particle size or higher dose of alumina catalyst can increase rate of mineralization of BPA but they cannot increase ultimate achievable removal of total organic carbon (TOC) beyond 90%. Adsorption of the reaction by-products on alumina catalyst surface plays an important role in ozonation of BPA (Keykavoos et al. 2013).

For a catalyst to reveal catalytic activity, adsorption of ozone and/or organic molecule on its surface has to take place. In heterogeneous catalytic ozonation, catalyst is in a solid form while the reaction may proceed in bulk water or on the surface of the catalyst. The catalytic ozonation is when the effect of ozonation in the presence of a catalyst is higher than a combined effect of adsorption on the catalyst surface and ozonation without a catalyst at the same pH. The formation of hydroxyl radicals is expected to be responsible for the catalytic effects. The catalytic effect is possible when at least one of the three conditions is fulfilled: ozone is adsorbed on the surface of the catalyst, BPA is adsorbed on the surface of the catalyst, or both, ozone and BPA are adsorbed on the catalyst surface. Among these metal oxides, nano-MnO_2 has been regarded as a promising catalyst for heterogeneous catalytic ozonation water treatment because of the peculiar properties including environmental friendliness and facile fabrication. The catalytic ozonation activity still needs to improve to meet the deteriorated water pollutions in nature environment. More recently, graphene oxide (RGO) has emerged as an excellent catalyst support. These hetero architectures are used as the highly efficient ozonation catalysts to eliminate BPA from water in presence of ozone. The high catalytic BPA degradation performance of α-MnO_2/RGO is attributed to the high surface area and abundant active sites in 3D sea urchin-like α-MnO_2. RGO in the nanocomposite accelerates the speed of electron transfer and improves the surface area of the catalyst. Moreover, the α-MnO_2/RGO catalyst is stable in the catalytic ozonation of BPA (Li et al. 2015b). As for the heterogeneous catalytic ozonation, the ozone adsorbed and decomposed on the surface of α-MnO_2 reacted with H_2O forming the active oxygen species, such as $HO^{\cdot}$. In the ozonation process, the $HO^{\cdot}$ has attacked phenyl groups in BPA to form p-hydroquinone and 4-(1-hydroxy-1-methylethyl)-phenol, then these intermediates were further oxidized to 4-hydroxyacetophenone and 4-benzoquinone (Katsumata et al. 2004). Ring-opening reaction occured at C–C bonds between the adjacent electron-donating substituents, forming organic acids such as oxalic acid, succinic acid and 2-hydroxypropanoic acid.

The degradation of BPA by ferrate(VI) oxidation can be divided into two possible pathways related to the bonds broken (Han et al. 2015). The first pathway is dehydroxylation, by which the two C–O bonds were cleaved and the two hydroxyl groups departed from BPA molecule, resulting in the formation of (2-methyl-1-phenyl-1-propenyl) benzene. Subsequent cleavage of the aromatic ring connecting C–C bond produces (1-methyl-vinyl) benzene and styrene, which each contains a single benzene ring. These intermediates are further oxidized to generate 2-phenylpropenal, 1,4-pentadien-3-ketone and maleic acid. The second pathway is initiated by the cleavage of the two phenyl groups in BPA and leads to the formation of p-isopropyl phenol, p-isopropenyl phenol and phenol. Two electron-donating hydroxyl groups in BPA molecule increases the electron density on each benzene ring and makes the C–C bond connecting the two benzene rings more vulnerable. These reaction products can continue to be oxidized by ferrate (VI), generating 2-phenyl isopropanol, 1-(4-methyl phenyl) ethanone, hydroquinone, benzoquinone, 1,4-pentadien-3-ketone and maleic acid (Li et al. 2008).

According to the radical (Weiss–Harber) mechanism, hydrogen peroxide (H_2O_2) can be decomposed into hydroxyl radical ($^{\cdot}OH$) by ferrous ions catalyst. $^{\cdot}OH$ has the second highest oxidant potential in nature (2.7 eV in an acidic solution) and can oxidize organic compounds in aqueous rapidly. The cost for this process could be too high in practical application because of the production of undisposable sludge, the continuous loss of the chemicals (mainly iron salt) and the narrow range of operation pH value (2.5–3.5). The heterogeneous Fenton agent was indicated as an ideal solution to these problems. Solid iron-based catalysts showed comparable catalytic performance to the conventional Fenton catalyst (Fe(II)); however, the leaching of iron generally leads to a significant deactivation. Au-based catalysts have been proven to be active for low-temperature oxidation, hydrogenation, water-gas shift, electrochemical reaction and so forth. Au/hydroxyapatite could be an alternative Fenton-like catalyst, which degrades H_2O_2 to $^{\cdot}OH$. Au atoms can be excellent electron-donor or -acceptor under certain circumstances; in particular, this property can be used for the production of $^{\cdot}OH$ by H_2O_2 decomposition. Au-Fenton catalysts have a great advantage compared to solid Fe-Fenton catalysts because Au is “inert” in nature and unlikely leaching under various reaction conditions. Au nanoparticles dispersed on various carbon supports were prepared, characterized and used as solid Fenton catalysts for the degradation of BPA under mild conditions. In particular, the origin of active sites over Au/C catalyst was explored using multi techniques (Navalon et al. 2011). Au/styrene-based activated carbon was effective Au-Fenton catalysts for the degradation of BPA by the production of $OH^{\cdot}$. It showed high activity and durability in a broad pH range (3–7) without extra energy input such as photon and electricity compared to other Au-Fenton catalysts.

Permanganate have a higher selectivity for BPA oxidation in natural water, which is generally rich in humic acid. The reaction pathways of BPA degradation by permanganate were similar to those of BPA destruction by ozonation (Zhang et al. 2013). Benzoic ring was the reaction site at the early stage of BPA oxidation by permanganate, ferrate, ozone, chlorine and hydroxyl radical. The co-existing solutes in the natural water had no negative effects on the BPA degradation; the presence of HCO_3^-, Fe(II), Fe(III) and humic acid could enhance the BPA removal.

3.2.2 Photochemical Oxidation

In the photocatalytic oxidation process, organic pollutants are destroyed in the presence of semiconductor photocatalysts (e.g. TiO_2 and ZnO), an energetic light source and an oxidising agent such as oxygen or air. The illumination of the photocatalytic surface with sufficient energy leads to the formation of a positive hole (h_v^+) in the valence band and an electron (e^-) in the conduction band. The positive hole oxidizes either pollutant directly or water to produce $^{\cdot}OH$ radicals, whereas the electron in the conduction band reduces the oxygen adsorbed on the catalyst (TiO_2). The photocatalytic activity of TiO_2 is dependent on surface and structural properties of the semiconductor such as crystal composition, surface area, particle size distribution, porosity, band gap and surface hydroxyl density (Ahmed et al. 2010). The photocatalyst titanium dioxide Degussa P-25 has been widely used in most of the experimental conditions; other catalyst powders, namely, Hombikat UV100, PC 500 and TTP were also used for degradation of toxic organic compounds. P-25 contains 75% anatase and 25% rutile with a specific BET surface area of 50 m^2/g and a primary particle size of 20 nm (Bahnemann et al. 2007). Since the discovery of photocatalytic effect on water splitting using TiO_2 electrode (Fujishima and Honda 1972), numerous researches have evolved to synthesis TiO_2 catalyst of different scale, characterize its physical properties and determine its photooxidation performances to the surface oriented nature of photocatalysis reaction. The TiO_2 catalyst in nanodimensions allows having a large surface area-to-volume ratio and can further promote the efficient charge separation and trapping at the physical surface. The light opaqueness of nanoscale TiO_2 catalysts can enhance the oxidation capability compared to the bulk TiO_2 catalysts. The effect of light intensity on the solar photocatalytic degradation of BPA in water with TiO_2 on sunny and cloudy days showed that efficiency of the degradation increase rapidly with increase in the light intensity up to 0.35 mW/cm^2, and then increased gradually (Kaneco et al. 2004).

The photocatalytic rate initially increases with catalyst loading and then decreases at high values because of light scattering and screening effects. The tendency toward agglomeration (particle-particle interaction) also increases at high solids concentration, resulting in a reduction in surface area available for light absorption and hence a drop in photocatalytic degradation rate. The number of active sites in solution will increase with catalyst loading, a point appears to be reached where light penetration is compromised because of excessive particle concentration. Two opposing phenomena results in an optimum catalyst loading for the photocatalytic reaction. Organic compounds in wastewater differ greatly in several parameters, particularly in their speciation behavior, solubility in water and hydrophobicity. While some compounds are uncharged at common pH conditions typical of natural water or wastewater, other compounds exhibit a wide variation in speciation (or charge) and physico-chemical properties. At a pH below its *pKa* value, an organic compound exists as a neutral species. Above this *pKa* value, an organic compound attains a negative charge. Some compounds can exist in positive, neutral, as well as negative forms in aqueous solution. This variation can also significantly influence their photocatalytic degradation behavior. The pH of the wastewater can vary significantly. The pH of an aquatic environment plays an important role on the photocatalytic degradation of organic contaminants since it determines the surface charge of the photocatalyst and the size of aggregates it forms (Singh et al. 2007). The surface charge of a photocatalyst and ionization or speciation of an organic pollutant can be profoundly affected by the solution pH. Inorganic anions, such as phosphate, sulphate, nitrate and chloride limited the performance of solar based photocatalysis. Bicarbonate in particular is detrimental to reactor performance as it acts as a hydroxyl radical scavenger. Long time experience with photocatalytic oxidation systems showed that humic substances in contaminated water can strongly adsorb titanium. The observed retardations of humic acids were related to the inhibition (surface deactivation), competition and light attenuation effects. The direct photodegradation of BPA showed that only 15% of the initial BPA was degraded after 180 min of treatment. The BPA absorption spectrum shows a maximum absorption peak at a wavelength of 275 nm (ε = 3290 L/mol/cm.)) and only 1.48% of the total energy emitted by the lamp corresponded to this wavelength, implying that the amount of radiation absorbed at 275 nm was inadequate to induce electron state transitions capable of transforming BPA, explaining the low degradation rate (Sánchez-Polo et al. 2013). The homogeneous UV/H_2O_2 system was adequate to degrade BPA, yielding a reaction rate constant of BPA with HO˙ radicals of $k_{HO^{\cdot}BPA} = 1.70 \pm 0.21 \times 10^{10}\, 1/(\mathrm{M\,s})$. The percentage of BPA degradation depended on the solution pH. The BPA degradation by-products detected with the UV/H_2O_2 system were monohydroxylated BPA and quinone of monohydroxylated BPA. The use of both UV/H_2O_2 and UV/Na_2CO_3 systems lead to the same degradation products in approximately the same proportion: a minor product consisting in monohydroxylated BPA and a major product assigned to the corresponding bis-quinonic derivative of the former. A possible mechanism leading to these derivatives should probably occur by the attack of a hydroxyl radical to BPA, followed by the capture of hydrogen radical by molecular oxygen reactions (Sánchez-Polo et al. 2013):

$$\mathrm{BPA} + \mathrm{OH}^{\cdot} \leftrightarrow \mathrm{BPAOH}^{\cdot}$$
$$\mathrm{BPAOH}^{\cdot} + \mathrm{O_2} \leftrightarrow \mathrm{BPA} - \mathrm{OH} + \mathrm{HOO}^{\cdot}$$

Interesting photocatalytic oxidation process is the combination of UV-A radiation and iron, just dissolved or as oxides. The efficiency of this oxidizing system can be improved by the presence of some carboxylic acids that in the presence of iron form carboxylate complexes that absorb UV-A radiation with high quantum yield to trigger radical chain mechanisms of oxidation (Rodriguez et al. 2009).

To understand the degradation of EDCs with existence of iron oxides and polycarboxylic acids in the natural environment, the photodegradation of BPA at the interface of iron oxides under UV illumination was conducted (Li et al. 2007). Four iron oxides were prepared by a hydrothermal process and then sintered at temperatures of 65, 280, 310 and 420 °C. A crystal structure of lepidocrocite (γ-FeOOH), a crystal structure of hematite (α-Fe_2O_3), and the mixed crystal structures of maghemite (γ-Fe_2O_3) and hematite were shown. The photodegradation of BPA depends strongly on the properties of iron oxides and oxalate and pH (Li et al. 2007). The properties of iron oxides influenced strongly the dependence of the BPA degradation on the oxalate concentration. The dependence of BPA degradation should be also attributable to the formation of the dissolved Fe-oxalate in the solution and the adsorbed Fe-oxalate on the surface of iron oxides, and a formation of hydrogen peroxide. Main factors affecting iron oxide dissolution are the properties of the overall system (temperature, UV light), the composition of the solution phase (pH, redox potential, concentration of acids, reductants and complexing agents) and the properties of the oxide (specific surface area, stoichiometry, crystal chemistry, crystal defects or guest ions). However, only the composition of the solution and the tendency of ions in solution to form complexes were considered important in mechanistic studies.

Peroxydisulfate ($S_2O_8^{2-}$) is a strong oxidant (potential 2.05 V) which has been used widely in the petroleum industry for the treatment of hydraulic fluids or as a reaction initiator. The reactions of peroxydisulfate are generally slow at normal temperature; the thermal or photochemical activated decomposition of $S_2O_8^{2-}$ ion to $SO_4^{\cdot-}$ radical has been proposed as a method of accelerating the process. The UV/$K_2S_2O_8$ system was more effective than UV/H_2O_2 to degrade BPA, achieving a higher percentage removal in a shorter time due to the generation of $SO_4^{\cdot-}$ and $HO^{\cdot}$ radicals (Khataee and Mirzajani 2010). Both removal of TOC and reduction of toxicity were the highest when using the UV/$K_2S_2O_8$ system. The rate constant was lower for the UV/$K_2S_2O_8$ system than for the UV/H_2O_2 system, indicating that the reaction is faster between BPA and $HO^{\cdot}$ radicals than between BPA and $SO_4^{\cdot-}$ radicals.

BPA oxidation varied as a function of the water type when the UV/H_2O_2, UV/$K_2S_2O_8$ and UV/Na_2CO_3 systems were used, decreasing in order: wastewater > surface water > ultrapure water. The high degradation of BPA in wastewater may be related to its large content of organic matter, which generates additional hydroxyl radicals from the photolysis of organic matter. With regard to their application, UV/$K_2S_2O_8$ was the most effective to remove BPA and TOC from the different water types.

A thermally activated persulfate oxidation process was used to treat aqueous BPA solution. When the treatment temperature increased (40–70 °C), a significant enhancement of BPA and its TOC removal were obtained (Olmez-Hanci et al. 2013). Acidic (pH 3.0) to closely neutral (pH 6.5) pHs were more favorable for BPA hot persulfate oxidation than alkaline pH (9.0 and 11.0). Several aromatic (benzaldehyde, *p*-isopropenyl phenol, 2,3-dimethyl benzoic acid, 3-hydroxy-4-methyl-benzoic acid) and some aliphatic (ethylene glycol monoformate and succinic acid) oxidation products were identified. The preliminary oxidation intermediates formed during the initial stages of treatment were found to be more toxic than BPA itself. The increase in acute toxicity was followed by an appreciable decrease (22%) at the end of the treatment period.

The combination of Fenton's reagent with UV light is called a photo-Fenton reaction. UV light irradiation enhances the efficiency of the Fenton process. The hydroxyl radical generated in the Fenton process is due to the iron catalysed decomposition of H_2O_2 as shown in the following:

$$Fe^{2+} + H_2O_2 \rightarrow {}^{\cdot}OH + Fe^{3+} + OH^-$$

In the Fenton process, hydrogen peroxide is added to wastewater in the presence of ferrous salts, generating species that are strongly oxidative with respect to organic compounds. $^{\cdot}OH$ is traditionally regarded as the key oxidizing species in Fenton processes. The Fenton process mechanism is quite complex and is described in detail with equations in the literature (Gallard and De Laat 2000). The formation of hydroxyl radical also occurs in the photo-Fenton:

$$Fe^{3+} + H_2O + UV \rightarrow {}^{\cdot}OH + Fe^{2+} + H^+$$

This is also attributed to the decomposition of the photoactive $Fe(OH)^{2+}$ which leads to the addition of the $HO^{\cdot}$ radicals:

$$Fe(OH)^{2+} + h\nu \rightarrow Fe^{2+} + {}^{\cdot}OH$$

$^{\cdot}OH$ radicals are responsible for oxidation; coagulation is ascribed to the formation of ferric complexes. The relative importance of oxidation and coagulation depends primarily on the H_2O_2/Fe^{2+} ratio. Chemical coagulation predominates at a lower H_2O_2/Fe^{2+} ratio, whereas chemical oxidation is dominant at higher H_2O_2/Fe^{2+} ratios (Neyens and Baeyens 2003). A huge molar excess of the H_2O_2 oxidant is required for complete mineralization of BPA, according to the reaction:

$$C_{15}H_{16}O_2 + 36\,H_2O_2 + 15\,CO_2 + 44\,H_2O$$

Pronounced sub-stoichiometric amounts of H_2O_2 oxidant were used to simulate economically viable processes and operation under not fully controlled conditions, as for example in in situ groundwater remediation. Aside from the most abundant benzenediols and the monohydroxylated BPA intermediate, which were detected as stable intermediates in earlier contributions, a wide array of aromatic products in the molecular weight range between 94 Da (phenol) and ~500 Da was identified (Poerschmann et al. 2010). The occurrence of aromatic intermediates larger than BPA, which typically share either a biphenyl- or a diphenylether structure, can be explained by oxidative coupling reactions of stabilized free radicals or by the addition of organoradicals (organocations) onto BPA molecules or benzenediols. The hydroxycyclohexadienyl radical of BPA was recognized to play central role in the degradation pathways. Ring opening products, including lactic, acetic and dicarboxylic acids, could be detected in addition to aromatic intermediates.

Heterogeneous Fenton oxidation was found to be an efficient and cost-effective method, in which iron catalyst is immobilized onto solid supports (i.e. activated carbon, carbon nanotube, zeolite or clay). The multi-walled carbon nanotube-supported magnetite (Fe_3O_4/MWCNT) as the heterogeneous catalyst for Fenton oxidation has been synthesized (Cleveland et al. 2014). Heterogeneous Fenton oxidation using Fe_3O_4 amended onto multi-walled carbon nanotube (Fe_3O_4/MWCNT) showed effective degradation of BPA in aqueous solutions. The intermediates and oxidation products produced by the Fenton oxidation of BPA at the H_2O_2/BPA ratio of 4 did not show any biological toxicity. The H_2O_2/BPA ratio of 4 was much lower than that for other heterogeneous Fenton of BPA (54 mol H_2O_2/mol BPA) and comparable to those for homogeneous Fenton of BPA (2–9 mol H_2O_2/mol BPA). The Fenton oxidation rate of BPA using the catalyst was 3.5 times higher at 50 °C than at 20 °C. The five cycles of the Fenton oxidation using the same catalyst resulted in the steady removal of BPA confirming high stability of the Fe_3O_4/MWCNT catalyst over the multiple Fenton reactions. The hydroxyl radical-driven oxidation was the major step among the multiple reactions in the heterogeneous Fenton oxidation.

The bimetallic oxides (composite oxides of iron and another metal element) have attracted an attention. It was found that $FeVO_4$ possessed higher catalytic activity for the activation of H_2O_2 than α-Fe_2O_3, Fe_3O_4 and γ-FeOOH, which is attributed to the simultaneous activation of H_2O_2 by both Fe(III) and V(V) in $FeVO_4$. It is noted that copper ions also exhibited Fenton-like behaviors. $CuFeO_2$ microparticles were used as a heterogeneous Fenton-like catalyst, and exhibited much higher catalytic activity towards the degradation of BPA in the presence of H_2O_2 compared with Cu_2O microparticles and Fe_3O_4 nanoparticles. The use of 1.0 g/L of $CuFeO_2$ microparticles induced nearly complete degradation of 0.1 mmol BPA/L in 120 min and 85% of TOC removal in 180 min in the presence of 20 mmol H_2O_2/L. The high catalytic activity was attributed to the synergistic effect of Cu(I)/Cu(II) and Fe(II)/Fe(III) redox pairs (Zhang et al. 2014).

Gamma radiation of water resulted in the formation of two reductive species, solvated electrons e^-_{aq} and hydrogen atoms H, and one oxidative species, hydroxyl radicals $^\cdot$OH, which might remove BPA in aqueous solution under different reaction condition

$$H_2O--\backslash/\backslash/\backslash/\backslash/\backslash/\rightarrow [0.28]\,{}^\cdot OH + [0.06]\,{}^\cdot H + [0.27]e^-_{aq}$$
$$+ [0.05]\,H_2 + [0.07]\,H_2O_2 + [0.27]\,H^+$$

The numbers in brackets are the G-values (yields) in μmol/Gy of energy. The unit of measure is joules per kilogram, and its name is gray (Gy).

To establish oxidizing conditions involving only the hydroxyl radical, the water solutions were saturated with N_2O, which quantitatively converts the hydrated electrons, e^-_{aq}, and hydrogen atoms, $^\cdot$H, to hydroxyl radicals (Peller et al. 2009):

$$e^-_{aq} + N_2O + H_2O \rightarrow N_2 + OH^- + {}^\cdot OH \quad k = 9.1 \times 10^9\,1/(M\,s),$$
$${}^\cdot H + N_2O \rightarrow {}^\cdot OH + N_2 \quad k = 2.1 \times 10^6\,1/(M\,s).$$

In deionized water, the hydroxyl radical is added to the aromatic ring of BPA (reaction rate $k = 6.9 \times 10^9$ (±0.2) 1/(M s)), to eventually form the prominent, long-lived, hydroxylated intermediate product. The initial hydroxyl radical addition reaction occurs, the hydroxylation is averted, and a different mechanistic pathway ensues. The removal constant for the hydroxyl radical reaction with BPA is 0.45 ± 0.04 μmol/kGy, corresponding to an overall degradation efficiency of 76%.

The BPA removal using gamma radiation depends on absorbed dose, initial BPA concentration, pH of solution, and dissolved oxygen (DO) (Guo et al. 2012). At an absorbed dose of 8 kGy, BPA removal percentages were 88.8, 93.2 and 97.1% at DO concentrations of 1.9, 6.8 and 17.2 mg/L, respectively. The higher BPA removal percentages and rate constants at the higher DO concentrations further indicated that ·OH radicals played an important role in BPA removal during gamma radiation. In solution containing 20 mg BPA/L, the following organic substances were present: 2% of $CHCl_3$, 2% of CH_3OH or 2% of fulvic acid. At DO concentration of 6.8 mg/L and pH of 7.1, BPA removal efficiency was 98.2% in presence of $CHCl_3$. Considering that BPA removal percentages increased in presence of $CHCl_3$ and decreased in presence of fulvic acid or CH_3OH, the corresponding rate constants were 0.503, 0.319 and 0.164 1/kGy. ·OH radical oxidation controlled BPA removal during gamma radiation. Mechanism of gamma radiation-induced BPA removal in aqueous solution was mainly attributed to ·OH radical oxidation (Guo et al. 2012).

Depending on the deployed state of the photocatalysts, photocatalytic reactors for water treatment can generally be classified into two main configurations: reactors with suspended photocatalyst particles and reactors with photocatalyst immobilised onto continuous inert carrier (Pozzo et al. 2000). The disparity between these two main configurations is that the first one requires an additional downstream separation unit for the recovery of photocatalyst particles while the latter permits a continuous operation.

Two major problems need to be solved for photocatalysis technology. One is how to design a photocatalytic reactor with high efficiency and large capacity; the other is how to reduce energy consumption of electric light source or use solar irradiation as the light source directly (Pozzo et al. 2000).

Since the advent of a microreactor in 1990s, it has been applied in the field of photocatalysis research immediately. Photocatalytic microreactor possesses the merits of high heat and mass transfer efficiency, large surface to volume ratio (>10,000 m^2/m^3), uniform illumination, low light attenuation and satisfactory catalytic effect. First disadvantage is that photocatalyst is usually immobilized onto some supporting materials, which will largely reduce interfacial surface area compared with suspended catalyst particles. Second, the characteristic size of microreactor makes the effective illumination area of a single channel quite small. Third, the processing capacity of a single channel in microreactor is small (<100 L/min). Krivec et al. (2015) improved the processing capacity by increasing the length or number of micro-channels. The macro-scale photocatalytic fluidized bed reactor provides a good mixing of reactant and catalyst, high heat and mass transfer rate, large effective illumination area, higher quantum efficiency than immobilized photocatalytic systems. But it also has some problems. First, the sizes of existing fluidized bed reactors are large (>10 mm), so the light attenuates seriously during penetration of solution or catalyst particles. This will lead to low catalytic efficiency, long reaction time (>1 h) or circulation of reaction solution, and no possibility to realize high output per unit reaction volume. Second, though the catalytic efficiency of nano-scale particles is well, the separation cost is quite high (Yang et al. 2016).

3.2.3 Ultrasonic Cavitation

Ultrasound is defined as any wave frequency that is greater than the upper limit of human hearing ability, i.e. it is at frequencies above 16 kHz (16,000 cycles/s). Ultrasonic wave consists of compression and expansion (rarefaction) cycles. Acoustic cavitations (tiny micro bubbles) are created when it reaches rarefaction cycle where a negative acoustic pressure is sufficiently large to pull the water molecules from each other (the critical molecular distance, R for water molecules is 10^{-8} m). As a result, 'voids' are created in the liquid. On the other hand, the acoustic pressure is positive during compression cycle of ultrasonic wave to push molecules apart. The phenomenon of cavitation consists of the repetitive and distinct three steps: formation (nucleation), rapid growth (expansion) during the compression/rarefaction cycles until they finally reach a critical size. After that, they start to undergo violent collapse (implosion) in the liquid. The collapse of cavitation bubbles near the micro-particle surface will generate high-speed micro jets of liquid in the order of 100 m/s. This will subsequently produce ultrasonic asymmetric shock wave upon implosion of cavitation bubbles which may cause direct erosion (damage) on the particle's surface and de-aggregation of particles to hinder agglomeration. Consequently, it will experience a decrease in particle size and an increase in reactive surface area available for the subsequent reaction. The severity of the cavitation erosion that cause pitting and cracking of the particle surface is strongly influenced by the solid particle size. Thermal dissociation of water and dissolved oxygen molecules in the cavities will convert them into reactive species such as $^{\cdot}OH$, hydrogen atoms $^{\cdot}H$, $^{\cdot}O$ atoms and hydroperoxyl radicals $^{\cdot}OOH$ and in the cooling phase, these radicals will recombine to form hydrogen peroxide and molecular hydrogen. The main degradation mechanisms of ultrasound are pyrolysis and radical reactions. The formation of $OH^{\cdot}$ radicals takes place inside the cavity in the presence of ultrasound by pyrolysis. The pyrolysis takes place inside the cavity and near the interface of the cavity and surrounding liquid at the time of collapse of the cavity in the presence of ultrasound. Pyrolysis takes place because of the very high temperatures reached during cavitation. These temperatures are in the range of 5200 and 1900 K in the cavity and interfacial region, respectively. The surfaces of the cavitation bubbles are hydrophobic; therefore, hydrophobic compounds can enter the bubbles and be pyrolyzed directly by the high temperatures and pressures. The hydrophilic compounds barely diffuse into the bubbles and are degraded indirectly by radicals in the bulk solution.

The hydrogen peroxide is one of the most effective additives to enhance the sonochemical degradation of aqueous pollutants (Mehrdad and Hashemzadeh 2010). Under ultrasound irradiation, hydrogen peroxide can dissociate into two hydroxyl radicals and acts as a secondary source of hydroxyl radicals. Without adding hydrogen peroxide, hydrogen peroxide can be generated by the recombination of two hydroxyl radicals generated by cavitations in aqueous solution and linearly accumulated. It is not clear whether generated hydrogen peroxide can be the secondary source of hydroxyl radicals, as added hydrogen peroxide acted, because the concentration of added hydrogen peroxide drastically decreases while generated hydrogen peroxide is linearly accumulated.

The degradation of phenol and BPA was enhanced by the addition of H_2O_2 during sonochemical processes (Lim et al. 2014). The effects of H_2O_2 on the sonochemical degradation of phenol were more effective than on BPA because of the hydrophobicity of BPA. The degradation of TOC was enhanced by the addition of H_2O_2 and the effects of H_2O_2 were quite significant in the BPA solution compared to phenol because some intermediates of BPA have high K_{ow} values and a greater probability to react with the hydroxyl radicals. The degradation ratios of TOC were lower than the degradation ratios of BPA because some intermediates cannot readily degrade during the sonochemical reaction. High concentrations of H_2O_2 can act as a radical scavenger (especially hydroxyl radicals). Therefore, the determination of an optimum H_2O_2 concentration is crucial. The chemical pathway and the rate of elimination depend on the volatility of the pollutant molecule. A molecule with a large Henry's constant will be incinerated inside the bubble of cavitation, while a nonvolatile molecule with a low Henry's constant will be oxidised by the $OH^{\cdot}$ ejected from the cavitation bubble.

The presence of dissolved oxygen is reported to improve sonochemical reactions. However, it is not necessary for water sonolysis because sonochemical oxidation can proceed in the presence of any gas such as air, nitrogen, argon and hydrogen. BPA ultrasonic degradation in water occurs mainly through reactions with hydroxyl radicals. Reaction rates are strongly affected by ultrasonic frequency, gas saturation and applied ultrasonic power. The rate of degradation BPA increased in the order $O_2 > Ar > air > N_2$ (Kitajima et al. 2006). A typical primary intermediate of BPA (2,3-dihydro-2-methylbenzofuran), formed during attack by $^{\cdot}OH$ radicals, was detected only in the presence of oxygen, revealing that a different reaction path was responsible for the enhancement of decomposition. It is known that the density of $^{\cdot}OH$ radicals is less in Ar than in O_2. For dissolved Ar, BPA decomposed directly into low-molecular weight compounds due to the high temperature and high pressure inside the cavitation bubbles. The intermediate still has one-half the molecular weight of BPA, and thus BPA may be decomposed mainly via radical reaction.

The process transforms BPA into biodegradable aliphatic acids that could be treated in a subsequent biological treatment. The BPA degradation upon ultrasonic action proceeds under different experimental conditions. The effects of saturating gas (oxygen, argon and air), BPA concentration (0.15–460 μmol/L), ultrasonic frequency (300–800 kHz) and power (20–80 W) were investigated. For a solution containing 118 μmol BPA/L, the best efficiency was obtained at 300 kHz, 80 W and with oxygen as saturating gas. Under these conditions, BPA can be readily eliminated by the ultrasound process ($\sim$90 min). However, even after long ultrasound irradiation times (9 h), more than 50% of chemical oxygen demand (COD) and 80% of TOC remained in the solution. Sonochemical BPA elimination is related to the local concentrations of both ˙OH radicals and substrate. The identified intermediates of the process were monohydroxylated bisphenol A, 4-isopropenylphenol, quinone of monohydroxylated bisphenol A, dihydroxylated bisphenol A, quinone of dihydroxylated bisphenol A, monohydroxylated-4-isopropenylphenol and 4-hydroxyacetophenone. The presence of these hydroxylated aromatic structures showed that the main ultrasonic BPA degradation pathway is related to the reaction of BPA with the ˙OH radical. After 2 h, these early products were converted into biodegradable aliphatic acids (Torres et al. 2008).

To increase the efficiency of BPA removal, experiments combined ultrasound with Fe(II) (100 μmol/L) and/or UV radiation (254 nm): ultrasound/UV, ultrasound/Fe(II) and ultrasound/UV/Fe(II). Both UV and Fe(II) induced hydrogen peroxide dissociation, leading to additional ˙OH radicals and complete COD and TOC removal. Thus, difficulties in obtaining mineralization of micropollutants like BPA through ultrasonic action alone can be overcome by the ultrasound/UV/Fe(II) combination. Moreover, this technique was found to be the most cost-effective one. So, the integrated ultrasound/UV/Fe(II) process was shown to be of interest for the treatment of BPA-contaminated wastewater (Torres et al. 2007).

Accumulation of hydrogen peroxide from ˙OH and ˙OOH radical recombination was observed under sonochemical degradation of BPA. Increases in BPA initial concentration slowed down the reaction rate but enhanced the sonochemical product yield. The sonochemical product yield for the decomposition of BPA at 300 kHz increased linearly with the BPA concentration (Gultekin and Ince 2008). The degree of BPA decay was fastest in the presence of air due to the formation of acids and excess radicals, and slowed down as the gas was replaced by argon and oxygen. The impact of large concentrations of hydroxyl radical scavengers such as carbonate and t-butanol decreased the rate of degradation, particularly when the scavenger concentration was considerably larger than that of BPA. Acidic to near-neutral pH and injection of air render faster decomposition than alkaline pH, and argon or oxygen injection. The lower rate of hydrogen peroxide accumulation in BPA solutions than in ultrapure deionized water is an evidence of OH radical-mediated oxidation as the major destruction pathway. The impact of carbonate and t-butanol as strong scavengers of OH radicals was insignificant at low concentrations of the scavengers, but significant when the fraction of scavenger to BPA was large. The degree of rate inhibition was much larger in the presence of large concentrations of t-butanol than carbonate, showing the significance of interfacial reactions in the destruction pathway of BPA.

Considering the coexistence of halomethanes and BPA in chlorinated drinking water, BPA was degraded under ultrasonic at the frequency of 20 kHz in the presence of CCl_4. The main intermediates resulting from BPA ultrasonic degradation were monohydroxylated BPA, phenol, 4-isopropenylphenol hydroquinone, 4-hydroxyacetophenone, 2-hydroxypropionic acid and glycerol (Guo and Feng 2009).

The sonochemical reactivity and the degradation of BPA and 17 α-ethinyl estradiol (EE2) in sonochemical reactor in the presence of a stainless steel wire mesh (SSWM), which acted as a catalyst, at a low frequency of 28 kHz with a contact time of 60 min were determined (Park et al. 2011). The formation of H_2O_2 was found to be dependent on the SSWM setting height. For a SSWM pore size of 1.0 mm, the sonochemical reactivity and the degradation of BPA and EE2, as a function of the SSWM area, followed the order: 128 cm^2 > 192 cm^2 > 384 cm^2 > no addition. This process can be accelerated by adding various kinds of catalyst. Excessive amount of solid particles present in the solution could result in the scattering of ultrasound waves that may decrease the rates of sonochemical degradation reactions.

Tourmaline is a natural siliceous mineral that is distributed worldwide. Its chemical formula can be expressed as $NaR_3Al_6(Si_6O_{18})(BO_3)_3(OH,F)_4$, where R represents Mg, Fe, Li or Mn, corresponding to different types of tourmaline. Tourmaline has been found to possess spontaneous and permanent poles, piezo-electric and pyroelectric characteristics and the continuous generation of anions and infrared rays. The presence of electric fields with an intensity of 10^6–10^7 V/m on the surface of micrometer-size particles is generally recognized. The extent of BPA removal was dependent on tourmaline concentration, H_2O_2 concentration and pH. The highest removal rate of BPA was achieved after 120 min under the following conditions: sonication (40 kHz, 500 W), tourmaline concentration 5.0 g/L, 50 mM H_2O_2 and pH 2.0. The synergism of homogeneous oxidation, a heterogeneous Fenton-like process and cavitation effects in response to ultrasonic irradiation was noted. The radical scavenger t-butanol inhibited BPA removal, which confirmed the fact that the generated hydroxyl radicals were the main reactive species mediating the oxidation. A possible reaction scheme of BPA degradation involving hydroxylation, oxidative skeletal rearrangement, demethylation and dehydration was proposed, and eight intermediates of BPA degradation in these processes were tentatively deduced (Yu et al. 2014).

Ultrasound has the advantages of being non-selective, creating no secondary pollution and is particularly effective in removing BPA. The ultrasonic irradiation alone is unable to provide high degradation efficiency of hydrophilic compounds which are usually hard to be decomposed by conventional cavitation phenomenon. Heterogeneous sonocatalytic degradation of BPA using different catalysts such as metal oxides is an alternative way to solve the problem of transforming BPA. The most common type of catalyst to be used is TiO_2 and its advantages are such as inexpensive, non-toxic, biologically stable and reusable. Loading of metal particles into TiO_2 catalysts either through doping or used as their composites can inhibit charge recombination between holes and electrons produced during the ultrasonic irradiation of TiO_2 to enhance the degradation efficiency. To enhance the ultrasonic-mediated dissociation of water molecules for production of highly reactive free $^{\cdot}OH$, the various additive such as H_2O_2, Fe(II), O_3 and different types of salt (to alter properties of solvent and rate of $^{\cdot}OH$ formation) into aqueous solution that contain recalcitrant organic contaminants have been used to improve the overall rate of degradation. The main drawbacks of this technique lie in the high cost of reagents (e.g. H_2O_2 or Fenton reagent). Sonochemical processes could be an effective alternative way for oxidation and completely mineralizing BPA.

3.2.4 *Electrochemical Degradation*

Among the many possible alternatives for treating BPA, electrochemical technology is quite promising due to its versatility. However, there are many important variables for the optimization of an electrochemical process; thus, the correct optimization of these variables and the proper choice of reactor and electrode material to be used in the process may lead to lower power consumption, thus making the electrochemical treatment economically viable. All these electrochemical treatments aim to efficiently mineralize organic pollutants using hydroxyl radicals, which are generated by the electrochemical oxidation of water. The electrode includes carbon, platinum, and dimensionally stable anodes, which are typically prepared by thermal deposition of a thin metal oxide layer (e.g., SnO_2, PbO_2 or IrO_2) on a base metal. Electrochemical oxidation mechanism is mainly based on the generation of the hydroxyl radicals at the electrode surface (Comninellis 1994). Two different types of mechanisms have been elaborated for electrochemical oxidation, such as direct and indirect oxidation methods (Rasalingam et al. 2014). In direct electrochemical oxidation, the degradation of BPA occurs directly over the anode material, where the hydroxyl radical ($^{\cdot}OH$) or the reactive oxygen species react with the organic compound. The pollutants are first adsorbed at the surface of the anode and are then degraded by an anodic electron transfer reaction as follows:

$$M + H_2O \rightarrow M(^{\cdot}OH) + H^+ + e^-.$$

In the indirect electrochemical oxidation, the BPA is treated in the bulk solution by oxidants, such as $^{\cdot}OH$, Cl_2, hypochlorite (ClO^-), peroxodisulfate ($S_2O_8^{2-}$), and ozone (O_3), which are electrochemically generated at the electrode surface. Even though high removal efficiencies are achieved by both the direct and indirect electrochemical oxidation processes, their effectiveness strongly depends on the treatment conditions including pH, current density, types and concentration of pollutants, supporting electrolyte, flow rate, electrode preparation method and nature of the electrode materials. Several electrode materials that include Pt, PbO_2, Ti–SnO_2, Ti/Pt, Ti/Pt–Ir, Ti/PbO_2, Ti/PdO–Co_3O_4, Ti/RhO–TiO_2, Ti coated oxides of Ru/Ir/Ta, IrO_2, Ti/RuO_2, SnO_2, PbO_2, and so forth, and boron-doped diamond (BDD) have been listed as efficient electrodes for the degradation of organics by electrochemical oxidation. Apart from these anode materials, graphite anodes are also considered as efficient materials for anodic oxidation of several organic pollutants. Particularly, the high oxygen, high electrocatalytic activity, chemical stability, long lifetime and cost-effectiveness have been credited for the high efficiency of these graphite electrodes.

When used as anodes, one important difference among these electrodes is their oxidizing power, which increases in the following order: TiO_2–RuO_2 < PbO_2 < SnO_2 < BDD. The adsorption (indicated by the adsorption enthalpy) of these radicals on each electrode surface is different, thus affecting the electrode's oxidizing power. The greater the adsorption enthalpy the lower the oxidizing power (Kapalka et al. 2008). In this sense, BDD electrodes are distinguished by their high oxidizing power, an important feature that is mainly a consequence of their high potential for O_2 evolution and the electrogeneration of hydroxyl radicals ($^{\cdot}OH$) weakly adsorbed on their surfaces:

$$BDD + H_2O \rightarrow BDD(^{\cdot}OH)_{ads} + H^+ + e^-$$

As pointed out by Mascia et al. (2007, 2010), the oxidation of organic molecules by hydroxyl radicals occurs in a small layer near the electrode surface, because these radicals are adsorbed on the electrode surface and are highly reactive; thus, due to mass transfer limitations, large-area electrodes would be required for the degradation of large volumes of wastewater within a reasonable time. Possible solutions to this problem are the use of flow reactors or/and the generation of inorganic oxidizing species such as active chlorine (Cl_2, HOCl, OCl^-), obtained by adding chloride ions to the solution.

For the electrochemical oxidation of BPA from aqueous solution, the graphite electrodes were used (Govindaraj et al. 2013). Diamond-based material was used as the anode and stainless steel was used as the cathode. COD removal of 78.3% was obtained when 0.05 M NaCl was used as the electrolyte at an initial pH of 5 and a current density of 12 mA/cm^2. The oxidation kinetic constants (k) using different electrolytes were found to be in the following order: $k$$Na_2SO_4$ ≈ $k$$Na_2CO_3$ > $k$$H_2SO_4$ > $k$$H_3PO_4$. However, the addition of chloride ion to the electrolyte solution caused a major change in the reaction kinetics, and the rate constants were found to be in the following order: $k$$H_2SO_4$ > $k$$Na_2SO_4$ > $k$$Na_2CO_3$. The difference in the reaction kinetics was obtained by the oxidation mediated by the chloride ions, which were formed at low pH values. Though, the concentration of chloride ions helped to rapidly increase the reaction rate; increment in the supporting electrolyte concentration also improved the reaction kinetics. The Ti/BDD, Ti/Sb–SnO_2, Ti/RuO_2 and Pt were tested and compared for their performance in anodic oxidation of BPA. At a lower current density of 10 mA/cm^2, the best result on BPA destruction and TOC removal was obtained with Ti/Sb-SnO_2 anode. At a higher current density of 50 mA/cm^2, the Ti/BDD anode showed the best performance in BPA degradation, and passivation of Ti/Sb–SnO_2 anode was found (Cui et al. 2009). Pt had a lower capability than Ti/BDD and Ti/Sb–SnO_2 for BPA removal, and Ti/RuO_2 had the lowest effectiveness in BPA degradation. BPA was attacked by $^{\cdot}OH$ to form hydroxylated BPA derivatives which were transformed to one-ring aromatic compounds through the isopropylidene bridge cleavage. The aromatic compounds then underwent ring breakage and produced aliphatic acids. Finally, the organic acids were mineralised by electrolysis to CO_2. Compared to Pt and Ti/RuO_2 anodes, Ti/Sb–SnO_2 and Ti/BDD anodes had a higher oxygen evolution potential and a higher anodic potential for BPA electrolysis under the same current conditions. A greater ·OH production was expected for Ti/Sb–SnO_2 and Ti/BDD anodes. However, in consideration of such potential problems as rapid passivation and structure deterioration for the Ti/Sb–SnO_2 anode, Ti/BDD is apparently a more promising type of anodes for effective electrochemical degradation of BPA.

A batch reactor is the most widely used electrochemical reactor (Pozzo et al. 2000). The batch reactor usually consists of a container with a magnetic stirrer and two stick electrodes. This kind of reactor has simple configuration and low construction cost, but exhibits poor mass transfer characteristics and low treatment efficiency. Other kinds of reactors were proposed, such as the filter-press reactor, one-compartment electrolytic flow cell with parallel plate electrodes and bipolar trickle tower reactor using Raschig ring shaped BDD electrodes. These reactors improve the mass transfer by means of increasing the ratio of electrode area to volume, and thus present good performance. In contrast to the batch reactors, continuous-flow reactors have good application prospects due to their larger throughput and highest ability in terms of the effluent quality. For a continuous-flow reactor, the hydraulic characteristics have an important effect on the treatment efficiency. Conventional single cell reactors operated in continuous mode may result in a significant back mixing due to the agitation of the released bubbles, leading to a decrease in treatment efficiency. Moreover, when treating low concentrated solutions, high energy consumption is usually required because direct electrochemical oxidation is inhibited at low concentration of the target; most of the applied electric energy is wasted for oxygen evolution.

Electrochemical treatment of BPA was carried out using an undivided flow through cell. Cobalt promoted PbO_2 and stainless steel were used as the anode and cathode, respectively. The surface area of the electrodes was 10 cm^2 and the interelectrode distance was 0.5 cm. The range of current densities was from 0 to 50 mA/cm^2. The flow rate was set at 25 mL/min, which corresponded to a 12 s hydraulic residence time in the cell (Korshin et al. 2006). Electrochemical treatment and conventional chlorination caused diethylstilbestrol (DES) and BPA to undergo a rapid degradation accompanied by the generation of low molecular weight chlorinated organic species indicative of the breakdown of DES and BPA. The identified compounds were predominated by chloroacetic acids (HAAs), but approximately 80% of the total organic halogen was comprised by unidentified species. For electrochemical degradation, the HAA yields were lower and HAAs were predominated by monochloroacetic acid (MCAA), while in the case of conventional chlorination, trichloroacetic acid was predominant and MCAA was virtually absent.

A one-compartment filter-press reactor with a BDD anode was used for analyzing the influence of volumetric flow rate, temperature, pH, current density and NaCl addition on the electrooxidation of BPA (Pereira et al. 2012). The performance of the Nb/BDD anode is comparable with a β-PbO_2 or a TiO_2–RuO_2 anode. The addition of chloride ions to the BPA solution leads to an enhanced oxidation performance irrespective of anode used; total COD removal is attained only when the Nb/BDD anode is used. The superior performance presented by the Nb/BDD anode, which may be accounted for by the high oxidation power of the hydroxyl radicals electro generated at its surface, allows concluding that this anode might be option for the remediation of effluents containing BPA.

3.3 Membrane Filtration

Although EDCs, including BPA, may be eliminated from wastewater through physical and biological processes, residual concentrations may be found in the effluent causing estrogenic effects (Clara et al. 2004; Wintgens et al. 2004; Sumpter and Johnson 2005). Among the various water and wastewater treatment technologies that are being applied with the aim of eliminating EDCs and guaranteeing complete protection for health and the environment, membrane systems have become significant in recent years. Compared to the conventional processes, the remarkable advantage of membranes is their high pollutant retention capacity, producing water of excellent quality including extremely low organic concentration and removal of microbes and viruses without chemical disinfection (Liang et al. 2015). In addition, scientific and technical progress in membrane technology and their extended usage have led to reduced costs, promoting the further use of membranes (Nicolaisen 2002). However, membrane filtration transfers BPA from one medium to another resulting in the requirement of further treatment or disposal (Liang et al. 2015). Thus, the necessity of synthesizing new multifunctional membrane which can retain and decompose BPA simultaneously and of treating the backwash water of membrane process is emphasized (Liang et al. 2015).

BPA retention in the membrane processes is mainly due to size exclusion, charge repulsion and adsorption that is expected to be driven both by hydrophobic adsorption and hydrogen bonding (Wu et al. 2010). Adsorption on the surface and in the inside structure of membranes is known to play a significant role in removal of hydrophobic compounds, such as BPA (Bing-zhi et al. 2010; Seyhi et al. 2012). On the other hand, it has been suggested that physical adsorption on membranes is insignificant in pollutant removal because sorption is reversible and there is a limited active surface on the membrane with a limited number of sorption sites (Sun et al. 2015). In addition, BPA retained by membrane can be released into permeate if the BPA concentration in the influent had an erratic behavior (Liang et al. 2015). According to Wu et al. (2010), the presence of NOM decreased the adsorption of BPA onto membrane due to its competitive adsorption onto the limited adsorption sites of the membrane.

BPA rejection by membranes ranges from 18% (Kimura et al. 2004) to >99.9% (Agenson et al. 2003). This wide range of rejection rates is due to the fact that there is a strong relationship between rejection rate and membrane type; for phenolic compounds in particular, there is a linear relationship between rejection efficiency and membrane molecular weight *cut-off* (Jung et al. 2007). For this reason, the rejection efficiency of EDCs, including BPA, decreases in the following order: reverse osmosis (RO), nanofiltration (NF), ultrafiltration (UF) and microfiltration

(MF). Among the various treatment options for BPA removal from solutions, NF and low pressure techniques (MF and UF) are worth using. Although NF is effective for EDC removal since it can separate organic micropollutants of low molecular weight (e.g. BPA of molecular mass of 228.29 g/mol), it needs high driving force for operation, which increases operating costs. Low-pressure membrane processes such as MF and UF also have a good performance for EDCs removal, which can be attributed to adsorption mechanism.

Among various options in membrane technology, NF is increasingly being used for the removal of micropollutants from drinking water and wastewater (Wintgens et al. 2002; Agenson et al. 2003; Kimura et al. 2003; Yoon et al. 2006; Jin et al. 2007). During NF of BPA solutions, rejection was related to a size exclusion mechanism and the ability of the membrane to adsorb BPA (Escalona et al. 2014). Some NF membrane results suggested that the rejection of uncharged organic molecules is mainly influenced by the sieving mechanism and that solute transport takes place by convection due to a pressure difference and by diffusion due to a concentration gradient across the membrane (Kosutic and Kunst 2002). Other authors consider that as NF membranes span the gap between UF and RO membranes, and the sorption-diffusion mechanism in a non-porous structure can contribute to the separation process (Zhang et al. 2006). When using NF membrane, over 90% rejection of BPA at the beginning of filtration was achieved due to adsorption of BPA onto the membrane structure. However, the observed retention coefficient decreased to around 50% when the membrane became saturated. Due to the adsorption of BPA onto the membrane structure, the membrane could be considered as a reservoir of EDCs and retained compounds could be released into the permeate. Another disadvantage is that NF itself does not destroy micropollutants. As a result, organic pollutants in the NF concentrate stream ultimately must be mineralized and/or transformed into less harmful compounds (Kim et al. 2008b). BPA rejection by NF is influenced by pH of the solution, which change the electrical charge of the BPA molecule. Nghiem et al. (2005) obtained a 40% removal of BPA at pH 6 and nearly 100% at pH 11. It is difficult to reject BPA under alkaline conditions when it is deprotonated (Liang et al. 2015). Bolong et al. (2010) chemically modified NF membrane so that the surface is negatively charged, which resulted in a twofold higher BPA removal efficiency compared to unmodified membrane.

Although MF and UF are relatively less effective than NF, they offer the advantage of operation at lower transmembrane pressures. This makes these processes worth investigating with the aim of improving their efficiency. In the case of MF and UF, the pore size of the membranes is considerably greater than the molecular size of BPA to be eliminated. The notable results of these technologies are mainly the results of the association of the organic micropollutant to the particulate material (Gómez et al. 2007). MF was able to remove 20–95% of BPA from drinking water (Bing-zhi et al. 2010). In the investigations of BPA removal from drinking water by hollow fiber MF membrane, high removal was obtained at the beginning of the filtration proving that adsorption played a major role in BPA removal. The removal efficiency dropped to around 20% when the membrane became saturated. As pH of solution approached to dissociation constants *pKa* (9.6–11.3) of BPA, BPA removal efficiency dropped significantly (Dong et al. 2010a). The UF of 100 μg BPA/L by membranes of pore sizes of 10 kDa, 6 kDa and 2 kDa gave the rejection efficiencies of 93.0, 88.9 and 97.7%, respectively (Dong et al. 2008). BPA retention was significantly decreased from 92% at pH 4.0–7.0 to 9.3% at pH 10.4. The initial BPA concentration had little influence on BPA retention. The hydrophobic adsorption and competitive adsorption for sites were reported to be major mechanisms for BPA rejection.

Typical configurations of membrane application in wastewater treatment are combined systems of MF, UF and RO (Côté et al. 2004), allowing for flexibility, which means that the quality of the permeate can be adjusted to different purposes. Moreover MF and/or UF act as pre-treatment methods for RO. Two modes of treatment were tested for the purification of macrofiltered wastewater containing 760 ± 18 μg BPA/L (Gómez et al. 2007). The first system consisted of an MF module and an UF module working in parallel, and the second of an RO module. The RO membranes gave better results for BPA than for other EDCs examined even though BPA has a smaller molecular size. This pointed out that mechanisms other than size exclusion influenced the rejection. The RO resulted in 85% BPA rejection, however, final concentration was above the levels detected in water samples with evident toxic activity (Aguayo et al. 2004). In the MF and UF, 9 and 18.5% of BPA, respectively, were not detected in the balancing suggesting that BPA remained in the membrane, associated with particulate material left behind after backwashing or perhaps with the membrane itself (Gómez et al. 2007). In the RO, unbalanced 12.5% of BPA was reported to be associated with the membrane.

Hydrophobic compounds, such as BPA, tend to strongly bind to hydrophobic materials like membranes. This could change the hydrophobicity of the membrane surface, which results in fouling leading to the shortening of filtration cycle and membrane life, thus increasing the operational cost (Escalona et al. 2014). Opinions differ about reasons for fouling: it may be caused mainly by relatively large colloids and soluble organic compounds ranging from 0.450 to 0.026 μm (Zheng et al. 2009) or by extracellular polymers excreted by bacteria (Lee et al. 2006). Colloids that are deposited on the membrane may additionally adsorb dissolved organic compounds, which affect the rejection of micropollutants (Andrade et al. 2014). When fouling occurs during filtration, particles present in the wastewater clog membrane pores and form filtration cake. This decreases the filtration flux rate whilst simultaneously improving retention of pollutants. The rejection of particles smaller than the membrane *cut-off* is possible because fouling lowers the nominal diameter to the so-called "effective diameter". This has been confirmed by Muthukumaran et al. (2011), who showed that fouling affects permeate flux more than pore size. The increased rejection of EDCs was attributed partly to micropollutant-NOM partitioning and subsequent NOM retention, and partly to the fouling layer acting as a second membrane (Jermann et al. 2009). Also, Boonyaroj et al. (2012) found that BPA rejection by a membrane covered with a layer of pollutant was 69%, whereas rejection by a clean membrane was only 23%. Similarly, the presence of humic acid favored BPA removal in UF (Dong et al.

2008). On the other hand, the presence of NOM caused negligible fouling of NF and MF membranes and did not modify the BPA retention of the membrane (Zhang et al. 2006; Dong et al. 2010a).

In the modules for BPA rejection, different types of membranes were used, particularly fabricated from organic materials such as polysulfone and polyamide (Nghiem et al. 2008) or polyvinylidenefluoride (Bing-zhi et al. 2010). Polyamide (PA) MF membranes were reported as those that rejected BPA from water via sustained sorption to higher extent than polyethersulfone, cellulose acetate, mixed nitrocellulose, polyester, regenerated cellulose and polycarbonate membranes (Han et al. 2013). This strong binding affinity resulted from the hydrogen bonding interactions between PA amide groups and the proton donating moieties of the target compound molecules. PA membrane exhibited a sorption capacity of 0.32 μg BPA/cm^2. When examining polysulphone membrane, Wu et al. (2010) found that the BPA rejection was notably influenced by pH; a very poor BPA removal was observed at pH approaching *pKa* of BPA.

The rejection efficiency of the NF membrane is influenced by the interaction between the membrane and target compounds. Thus, there were some studies on evaluation of the ability of NF membranes to reject different EDCs, in relation to their physico-chemical properties (Yoon et al. 2006, 2007). In addition, an alternative strategy to improve rejection of EDCs is to modify surfaces of existing NF membranes. The main objective of membrane surface modifications was to reduce or prevent membrane fouling by NOM or microorganisms (Terada et al. 2005). In the studies by Kim et al. (2008a), a polyamide NF membrane was chemically modified to improve its rejection capacity for selected EDCs, including BPA. The membrane surface was altered by graft polymerization, cross-linking of grafted polymer chains and substitution of functional groups. BPA rejection was increased from 74 to $\geq$95% after graft polymerization of the membrane that increased the hydrophilicity and negative surface charge of the membrane. BPA rejection was further improved after cross-linking of grafted polymer chains with ethylene diamine; however, the water flux with this cross-linked membrane decreased due to the increased hydraulic resistance compared to the membrane after graft polymerization.

References

Aditya D, Rohan P, Suresh G (2011) Nano-adsorbents for wastewater treatment: a review. Res J Chem Environ 15:1033–1040

Agenson KO, Oh JI, Urase T (2003) Retention of a wide variety of organic pollutants by different nanofiltration/reverse osmosis membranes: controlling parameters of process. J Membr Sci 225:91–103

Aguayo S, Muñoz MJ, Torre A et al (2004) Identification of organic compounds and ecotoxicological assessment of sewage treatment plants (STP) effluents. Sci Total Environ 328:69–81

Ahmed S, Rasul MG, Martens WN et al (2010) Heterogeneous photocatalytic degradation of phenols in wastewater: a review on current status and developments. Desalination 261:3–18

Andrade LH, Mendes FDS, Espindola JC et al (2014) Nanofiltration as tertiary treatment for the reuse of dairy wastewater treated by membrane bioreactor. Sep Purif Technol 126:21–29

Bahnemann W, Muneer M, Haque MM (2007) Titanium dioxide-mediated photocatalysed degradation of few selected organic pollutants in aqueous suspensions. Catal Today 124:133–148

Bautista-Toledo I, Ferro-García MA, Rivera-Utrilla J et al (2005) Bisphenol A removal from water by activated carbon. Effects of carbon characteristics and solution chemistry. Environ Sci Technol 39:6246–6250

Beltran FJ, Rivas FJ, Montero-de-Espinosa R (2005) Iron type catalysts for the ozonation of oxalic acid in water. Water Res 39:3553–3564

Bing-zhi D, Hua-qiang C, Lin W et al (2010) The removal of bisphenol A by hollow fiber microfiltration membrane. Desalination 250:693–697

Bolong N, Ismail AF, Salim MR et al (2010) Negatively charged polyethersulfone hollow fiber nanofiltration membrane for the removal of bisphenol A from wastewater. Sep Purif Technol 73:92–99

Boonyaroj V, Chiemchaisri C, Chiemchaisri W et al (2012) Toxic organic micro-pollutants removal mechanisms in long-term operated membrane bioreactor treating municipal solid waste leachate. Bioresour Technol 113:174–180

Clara M, Strenn B, Saracevic E et al (2004) Adsorption of bisphenol-A, 17beta-estradiole and 17alpha-ethinylestradiole to sewage sludge. Chemosphere 56:843–851

Cleveland V, Bingham JP, Kan E (2014) Heterogeneous Fenton degradation of bisphenol A by carbon nanotube-supported Fe_3O_4. Sep Purif Technol 133:388–395

Comninellis C (1994) Electrocatalysis in the electrochemical conversion/combustion of organic pollutants for waste-water treatment. Electrochim Acta 39:1857–1862

Côté P, Masini M, Mourato D (2004) Comparison of membrane options for water reuse and reclamation. Desalination 167:1–11

Cui YH, Li XY, Chen G (2009) Electrochemical degradation of bisphenol A on different anodes. Water Res 43:1968–1976

Dai HJ, Hafner JH, Rinzler AG et al (1996) Nanotubes as nanoprobes in scanning probe microscopy. Nature 384:147–150

Deborde M, Rabouan S, Duguet JP et al (2005) Kinetics of aqueous ozone induced oxidation of some endocrine disruptors. Environ Sci Technol 39:6086–6092

Deborde M, Rabouan S, Mazellier P et al (2008) Oxidation of bisphenol A by ozone in aqueous solution. Water Res 42:4299–4308

Dong B, Wang L, Gao N (2008) The removal of BPA by ultrafiltration. Desalination 221:312–317

Dong B, Chu H, Wang L et al (2010a) The removal of BPA by hollow fiber microfiltration membrane. Desalination 250:693–697

Dong Y, Wu D, Chen X et al (2010b) Adsorption of bisphenol A from water by surfactant-modified zeolite. J Colloid Interface Sci 348:585–590

Escalona I, Fortuny A, Stüber F et al (2014) Fenton coupled with nanofiltration for elimination of bisphenol A. Desalination 345:77–84

Fujishima A, Honda K (1972) Electrochemical photolysis of water at a semiconductor electrode. Nature 238:37–38

Gallard H, De Laat J (2000) Kinetic modelling of $Fe(III)/H_2O_2$ oxidation reactions in dilute aqueous solution using atrazine as a model organic compound. Water Res 34:3107–3116

Garoma T, Matsumoto S (2009) Ozonation of aqueous solution containing bisphenol A: effect of operational parameters. J Hazard Mater 167:1185–1191

Ghosh S, Badruddoza AZM, Hidajat K et al (2013) Adsorptive removal of emerging contaminants from water using superparamagnetic Fe_3O_4 nanoparticles bearing aminated b-cyclodextrin. J Environ Chem Eng 1:122–130

Gómez M, Garralón G, Plaza F et al (2007) Rejection of endocrine disrupting compounds (bisphenol A, bisphenol F and triethyleneglycol dimethacrylate) by membrane technologies. Desalination 212:79–91

Govindaraj M, Rathinam R, Sukumar C et al (2013) Electrochemical oxidation of bisphenol-A from aqueous solution using graphite electrodes. Environ Technol 34:503–511

Gultekin I, Ince NH (2008) Ultrasonic destruction of bisphenol-A: the operating parameters. Ultrason Sonochem 15:524–529

Guo Z, Feng R (2009) Ultrasonic irradiation-induced degradation of low-concentration bisphenol A in aqueous solution. J Hazard Mater 163:855–860

Guo W, Hu W, Pan J et al (2011) Selective adsorption and separation of BPA from aqueous solution using novel molecularly imprinted polymers based on kaolinite/Fe_3O_4 composites. Chem Eng J 171:603–611

Guo Z, Dong Q, Hea D et al (2012) Gamma radiation for treatment of bisphenol A solution in presence of different additives. Chem Eng J 183:10–14

Han J, Meng S, Dong Y et al (2013) Capturing hormones and bisphenol A from water via sustained hydrogen bond driven sorption in polyamide microfiltration membranes. Water Res 47:197–208

Han Q, Wang H, Dong W et al (2015) Degradation of bisphenol A by ferrate(VI) oxidation: kinetics, products and toxicity assessment. Chem Eng J 262:34–40

Jermann D, Pronk W, Boller M et al (2009) The role of NOM fouling for the retention of estradiol and ibuprofen during ultrafiltration. J Membr Sci 329:75–84

Jin X, Hu JY, Ong SL (2007) Influence of dissolved organic matter on estrone removal by NF membranes and the role of their structures. Water Res 41:3077–3088

Joseph L, Heo J, Park YG et al (2011) Adsorption of bisphenol A and 17α-ethinyl estradiol on single walled carbon nanotubes from seawater and brackish water. Desalination 281:68–74

Jung YJ, Kiso Y, Park HJ et al (2007) Rejection properties of NF membranes for alkylphenols. Desalination 202:278–285

Kaneco S, Rahman MA, Suzuki T et al (2004) Optimization of solar photocatalytic degradation conditions of bisphenol A in water using titanium dioxide. J Photochem Photobiol A Chem 163:419–424

Kapalka A, Foti G, Comninellis C (2008) Kinetic modelling of the electrochemical mineralization of organic pollutants for wastewater treatment. J Appl Electrochem 38:7–16

Katsumata H, Kawabe S, Kaneco S et al (2004) Degradation of bisphenol A in water by the photo-Fenton reaction. J Photochem Photobiol Chem 162:297–305

Keykavoos R, Mankidy R, Mab H et al (2013) Mineralization of bisphenol A by catalytic ozonation over alumina. Separ Purif Technol 107:310–317

Khataee AR, Mirzajani O (2010) UV/peroxydisulfate oxidation of C.I. Basic Blue 3: modeling of key factors by artificial neural network. Desalination 251:64–69

Kim JH, Park PK, Lee CH et al (2008a) Surface modification of nanofiltration membranes to improve the removal of organic micro-pollutants (EDCs and PhACs) in drinking water treatment: graft polymerization and cross-linking followed by functional group substitution. J Membr Sci 321:190–198

Kim JH, Park PK, Lee CH et al (2008b) A novel hybrid system for the removal of endocrine disrupting chemicals: nanofiltration and homogeneous catalytic oxidation. J Membr Sci 312:66–75

Kim IT, Nunnery G, Jacob K et al (2010) Synthesis, characterization, and alignment of magnetic carbon nanotubes tethered with maghemite nanoparticles. J Phys Chem C 114:6944–6951

Kimura K, Amy G, Drewes J et al (2003) Rejection of organic micropollutants (disinfection by-products, endocrine disrupting compounds, and pharmaceutically active compounds) by NF/RO membranes. J Membr Sci 227:113–121

Kimura K, Toshima S, Amy G et al (2004) Rejection of neutral endocrine disrupting compounds (EDCs) and pharmaceutical active compounds (PhACs) by RO membranes. J Membr Sci 245:71–78

Kitajima M, Hatanaka S, Hayashi S (2006) Mechanism of O_2-accelerated sonolysis of bisphenol A. Ultrasonics 44:371–373

Kondratyuk P, Yates JT (2007) Molecular views of physical adsorption inside and outside of single-wall carbon nanotubes. Acc Chem Res 40:995–1004

Kono H, Onishi K, Nakamura T (2013) Characterization and bisphenol A adsorption capacity of β-cyclodextrin–carboxymethylcellulose-based hydrogels. Carbohydr Polym 98:784–792

Korshin GV, Kim J, Gan L (2006) Comparative study of reactions of endocrine disruptors bisphenol A and diethylstilbestrol in electrochemical treatment and chlorination. Water Res 40:1070–1078

Kosutic K, Kunst B (2002) Removal of organics from aqueous solutions by commercial RO and NF membranes of characterized porosities. Desalination 142:47–56

Krivec M, Pohar A, Likozar B et al (2015) Hydrodynamics, mass transfer, and photocatalytic phenol selective oxidation reaction kinetics in a fixed TiO_2 microreactor. AIChE J 61:572–581

Kruk M, Jaroniec M, Ryoo R et al (2000) Characterization of ordered mesoporous carbons synthesized using MCM-48 silicas as templates. J Phys Chem B 104:7960–7968

Kusvuran E, Yildirim D (2013) Degradation of bisphenol A by ozonation and determination of degradation intermediates by gas chromatography–mass spectrometry and liquid chromatography–mass spectrometry. Chem Eng J 220:6–14

Lee WN, Kang IJ, Lee CK (2006) Factors affecting filtration characteristics in membrane coupled moving bed biofilm reactor. Water Res 40:1827–1835

Li FB, Li XZ, Li XM et al (2007) Heterogeneous photodegradation of bisphenol A with iron oxides and oxalate in aqueous solution. J Colloid Interface Sci 311:481–490

Li C, Li XZ, Graham N et al (2008) The aqueous degradation of bisphenol A and steroid estrogens by ferrate. Water Res 42:109–120

Li X, Chen S, Fan X et al (2015a) Adsorption of ciprofloxacin, bisphenol and 2-chlorophenol on electrospun carbon nanofibers: in comparison with powder activated carbon. J Colloid Interface Sci 447:120–127

Li G, Lu Y, Lu C et al (2015b) Efficient catalytic ozonation of bisphenol-A over reduced graphene oxide modified sea urchin-like α-MnO2 architectures. J Hazard Mater 294:201–208

Liang L, Zhang J, Feng P et al (2015) Occurrence of bisphenol A in surface and drinking waters and its physicochemical removal technologies. Front Environ Sci Eng 9:16–38

Libbrecht W, Vandaele K, De Buysser K et al (2015) Tuning the pore geometry of ordered mesoporous carbons for enhanced adsorption of bisphenol-A. Materials 8:1652–1665

Lim M, Son Y, Khim J (2014) The effects of hydrogen peroxide on the sonochemical degradation of phenol and bisphenol A. Ultrason Sonochem 21:1976–1981

Liu G, Ma J, Li X et al (2009) Adsorption of bisphenol A from aqueous solution onto activated carbons with different modification treatments. J Hazard Mater 164:1275–1280

Mascia M, Vacca A, Palmas S et al (2007) Kinetics of the electrochemical oxidation of organic compounds at BDD anodes: modelling of surface reactions. J Appl Electrochem 37:71–76

Mascia M, Vacca A, Polcaro AM et al (2010) Electrochemical treatment of phenolic waters in presence of chloride with boron-doped diamond (BDD) anodes: experimental study and mathematical model. J Hazard Mater 174:314–322

Mehrdad A, Hashemzadeh R (2010) Ultrasonic degradation of rhodamine B in the presence of hydrogen peroxide and some metal oxide. Ultrason Sonochem 17:168–172

Mohmood I, Lopes CB, Lopes I et al (2013) Nanoscale materials and their use in water contaminants removal—a review. Environ Sci Pollut Res 20:1239–1260

Muthukumaran S, Nguyen DA, Baskaran K (2011) Performance evaluation of different ultrafiltration membranes for the reclamation and reuse of secondary effluent. Desalination 279:383–389

Mvula E, von Sonntag C (2003) Ozonolysis of phenols in aqueous solution. Org Biomol Chem 1:1749–1756

Nakanishi A, Tamai M, Kawasaki N et al (2002) Adsorption characteristics of bisphenol A onto carbonaceous materials produced from wood chips as organic waste. J Colloid Interface Sci 252:393–396

Navalon S, de Miguel M, Martin R et al (2011) Enhancement of the catalytic activity of supported gold nanoparticles for the Fenton reaction by light. J Am Chem Soc 133:2218–2226

Neyens E, Baeyens J (2003) A review of classic Fenton's peroxidation as an advanced oxidation technique. J Hazard Mater 98:33–50

Nghiem LD, Schäfer AI, Elimelech M (2005) Nanofiltration of hormone mimicking trace organic contaminants. Separ Sci Technol 40:2633–2649

Nghiem LD, Vogel D, Khan S (2008) Characterising humic acid fouling of nanofiltration membranes using bisphenol A as a molecular indicator. Water Res 42:4049–4058

Nicolaisen B (2002) Developments in membrane technology for water treatment. Desalination 153:355–360

Ning B, Graham N, Zhang Y et al (2007) Degradation of endocrine disrupting chemicals by ozone/AOPs. Ozone Sci Eng 29:153–176

Olmez-Hanci T, Arslan-Alaton I, Genc-Istanbul B (2013) Bisphenol A treatment by the hot persulfate process: oxidation products and acute toxicity. J Hazard Mater 263:283–290

Pan B, Xing B (2008) Adsorption mechanisms of organic chemicals on carbon nanotubes. Environ Sci Technol 42:9005–9013

Park JS, Her N, Oh J et al (2011) Sonocatalytic degradation of bisphenol A and 17 -ethinyl estradiol in the presence of stainless steel wire mesh catalyst in aqueous solution. Sep Purif Technol 78:228–236

Peller JR, Stephen C, Mezyk P et al (2009) Bisphenol A reactions with hydroxyl radicals: diverse pathways determined between deionized water and tertiary treated wastewater solutions. Res Chem Intermed 35:21–34

Pereira GF, Rocha-Filho RC, Bocchi N et al (2012) Electrochemical degradation of bisphenol A using a flow reactor with a boron-doped diamond anode. Chem Eng J 198–199:282–288

Poerschmann J, Trommler U, Górecki T (2010) Aromatic intermediate formation during oxidative degradation of bisphenol A by homogeneous sub-stoichiometric Fenton reaction. Chemosphere 79:975–986

Pozzo RL, Giombi JL, Baltanas MA et al (2000) The performance in a fluidized bed reactor of photocatalysts immobilized onto inert supports. Catal Today 62:175–187

Rasalingam S, Peng R, Koodali RT (2014) Removal of hazardous pollutants from wastewaters: applications of TiO_2–SiO_2 mixed oxide materials. J Nanomater. https://doi.org/10.1155/2014/617405

Ren X, Chena C, Nagatsu M et al (2011) Carbon nanotubes as adsorbents in environmental pollution management: a review. Chem Eng J 170:395–410

Rodriguez EM, Nunez B, Fernandez G et al (2009) Effects of some carboxylic acids on the Fe(III)/UVA photocatalytic oxidation of muconic acid in water. Appl Catal B Environ 89:214–222

Rokhina EV, Virkutyte J (2010) Advanced catalytic oxidation of emerging micropollutants. In: Virkutyte J, Varma RS, Jegatheesan V (eds) Treatment of micropollutants in water and wastewater. Integrated environmental technology series. IWA Publishing, London, pp 360–424

Sánchez-Polo M, Abdeldaiem MM, Ocampo-Pérez R et al (2013) Comparative study of the photodegradation of bisphenol A by $HO^{\cdot}$, $SO_4^{\cdot-}$ and $CO_3^{\cdot-}/HCO_3^{\cdot}$ radicals in aqueous phase. Sci Total Environ 463–464:423–431

Sauleda R, Brillas E (2001) Mineralization of aniline and 4-chlorophenol in acidic solution by ozonation catalyzed with Fe^{2+} and UVA light. Appl Catal B Environ 29:135–145

Seyhi B, Drogui P, Buelna G et al (2012) Removal of bisphenol-A from spiked synthetic effluents using an immersed membrane activated sludge process. Sep Purif Technol 87:101–109

Singh HK, Saquib M, Haque M et al (2007) Titanium dioxide mediated photocatalysed degradation of phenoxyacetic acid and 2,4,5-trichlorophenoxyacetic acid, in aqueous suspensions. J Mol Catal A: Chem 264:66–72

Sumpter JP, Johnson AC (2005) Lessons from endocrine disruption and their application to other issues concerning trace organics in the aquatic environment. Environ Sci Technol 39:4321–4322

Sun X, Wang C, Li Y et al (2015) Treatment of phenolic wastewater by combined UF and NF/RO processes. Desalination 355:68–74

Terada A, Yuasa A, Tsuneda S et al (2005) Elucidation of dominant effect on initial bacterial adhesion onto polymer surfaces prepared by radiation-induced graft polymerization. Colloid Surf B 43:99–107

Torres RA, Abdelmalek F, Combet E et al (2007) A comparative study of ultrasonic cavitation and Fenton's reagent for bisphenol A degradation in deionised and natural waters. J Hazard Mater 146:546–551

Torres RA, Petrier C, Combet E et al (2008) Ultrasonic cavitation applied to the treatment of bisphenol A. Effect of sonochemical parameters and analysis of BPA by-products. Ultrason Sonochem 15:605–611

Tsai WT, Hsu HC, Su TY et al (2006a) Adsorption characteristics of bisphenol-A in aqueous solutions onto hydrophobic zeolite. J Colloid Interface Sci 299:513–519

Tsai WT, Lai CW, Su TY (2006b) Adsorption of bisphenol-A from aqueous solution onto minerals and carbon adsorbents. J Hazard Mater B 134:169–175

Wintgens T, Gallenkemper M, Melin T (2002) Endocrine disruptor removal from wastewater using membrane bioreactor and nanofiltration technology. Desalination 146:387–391

Wintgens T, Gallenkemper M, Melin T (2004) Removal of endocrine disrupting compounds with membrane processes in wastewater treatment and reuse. Water Sci Technol 50:1–8

Wu S, Dong B, Huang Y (2010) Adsorption of BPA by polysulphone membrane. Desalination 253:22–29

Yamamoto Y, Niki E, Shiokawa H et al (1979) Ozonation of organic compounds. Ozonation of phenol in water. J Organ Chem 44:2137–2142

Yang Z, Liu M, Lin C (2016) Photocatalytic activity and scale-up effect in liquid–solid mini-fluidized bed reactor. Chem Eng J 291:254–268

Youn SC, Jung D, Ko YK et al (2009) Vertical alignment of arbon nanotubes using the magneto-evaporation method. J Am Chem Soc 131:742–748

Yoon Y, Westerhoff P, Snyder SA et al (2006) Nanofiltration and ultrafiltration of endocrine disrupting compounds, pharmaceuticals and personal care products. J Membr Sci 270:88–100

Yoon Y, Westerhoff P, Snyder SA et al (2007) Removal of endocrine disrupting compounds and pharmaceuticals by nanofiltration and ultrafiltration membranes. Desalination 202:16–23

Yu L, Wang C, Ren X et al (2014) Catalytic oxidative degradation of bisphenol A using an ultrasonic-assisted tourmaline-based system: Influence factors and mechanism study. Chem Eng J 252:346–354

Zhang Y, Causserand C, Aimar P et al (2006) Removal of BPA by a nanofiltration membrane in view of drinking water production. Water Res 40:3793–3799

Zhang SJ, Shao T, Bekaroglu SSK et al (2009) The impacts of aggregation and surface chemistry of carbon nanotubes on the adsorption of synthetic organic compounds. Environ Sci Technol 43:5719–5725

Zhang J, Sun B, Guan X (2013) Oxidative removal of bisphenol A by permanganate: Kinetics, pathways and influences of co-existing chemicals. Sep Purif Technol 107:48–53

Zhang X, Ding Y, Tang H et al (2014) Degradation of bisphenol A by hydrogen peroxide activated with $CuFeO_2$ microparticles as heterogeneous Fenton-like catalyst: efficiency, stability and mechanism. Chem Eng J 236:251–262

Zheng X, Ernst M, Jekel M (2009) Identification and quantification of major organic foulants in treated domestic wastewater affecting filterability in dead-end ultrafiltration. Water Res 43:238–244

Zheng S, Sun Z, Park Y et al (2013) Removal of bisphenol A from wastewater by Ca-montmorillonite modified with selected surfactants. Chem Eng J 234:416–422

Zhu Y, Murali S, Stoller MD et al (2011) Carbon-based supercapacitors produced by activation of graphene. Science 332:1537–1541

Chapter 4
Microbial Biodegradation and Metabolism of BPA

Microorganisms play an important role in BPA removal. The BPA molecule can be readily metabolized by many microbial communities and BPA-degrading strains have been isolated from water, soil and biomass from wastewater treatment systems. BPA degradation in the environment is mainly due to bacterial metabolism, though, the activities of fungi and algae in BPA degradation are also discussed. The metabolites produced during degradation of BPA under aerobic conditions have been exhaustively studied and several BPA degradation pathways have been proposed. Some information is also presented on the enzymes and genes that are involved in BPA degradation. The role of species composition and adaptation of the microbial community to BPA removal is discussed, as well as environmental factors that may influence the efficiency of BPA degradation. A summary of this information will help readers better understand the fate of BPA in the environment, how BPA degradation by different groups of microorganisms proceeds and finally, how to design treatment lines to ensure efficient BPA removal.

4.1 BPA Degradation by Bacteria

BPA degradation is best studied for bacteria and a review of some bacterial strains which are able to degrade BPA is presented in Table 4.1. Although bacteria capable of BPA degradation are widely distributed in different environments, research have shown that bacterial strains differ significantly in the ability to efficient BPA degradation and that the number of bacteria with a high BPA-degradation potential is limited. Kang and Kondo (2002a) found out that out of 11 bacterial strains isolated from river waters, 10 possessed BPA degradation potential, but only two of them showed biodegradability over 90%. In BPA-rich samples, bacteria belonging to *Sphingomonas* sp. are most often identified since they are able to metabolize BPA as the sole source of carbon and energy (Matsumura et al. 2009). For example, *Sphingomonas bisphenolicum* strain AO1 was able to degrade 115 mg BPA/L

M. ZIELIŃSKA et al., *Bisphenol A Removal from Water and Wastewater*,
https://doi.org/10.1007/978-3-319-92361-1_4

Table 4.1 A review of exemplary bacterial strains which are able to degrade BPA

Strain	Source	References
Achromobacter xylosoxidans strain B-16	Compost leachate of municipal solid waste	Zhang et al. (2007)
Aeromonas hydrophilia	Water	Gulnaz and Dincer (2009)
Alcaligenes sp. strain OIT7	Soil	Matsumura et al. (2009)
Bacillus sp. NO15, NO13, YA27	Soil	Matsumura et al. (2009)
Bacillus sp. strain GZB	Sediment/soil	Li et al. (2012)
Bacillus sp. strain KU3	Seawater sediments	Kamaraj et al. (2014)
Bordetella sp. strain OS17	Soil	Matsumura et al. (2009)
Cupriavidus basilensis strain JF1	Wastewater treatment system	Fischer et al. (2010)
Enterobacter sp. strains HI9 and HA18	Soil	Matsumura et al. (2009)
Enterobacter gergoviae strain BYK-1	Wastewater treatment system	Badiefar et al. (2015)
Klebsiella sp. strains NE2, SU3 and SU5	Soil	Matsumura et al. (2009)
Klebsiella pneumoniae strain BYK-9	Wastewater treatment system	Badiefar et al. (2015)
Methylomonas clara	River sediments	Peng et al. (2015)
Nitrosomonas europaea	Pure culture	Roh et al. (2009)
Novosphingobium sp. strain TYA-1	Rhizosphere of plant	Toyama et al. (2009)
Pandoraea sp. strain HYO6	Soil	Matsumura et al. (2009)
Pseudomonas knackmussii	River sediments	Peng et al. (2015)
Pseudomonas paucimobilis strain FJ-4	Wastewater treatment plant	Ike et al. (1995)
Pseudomonas putida strain KA5	Water	Kang and Kondo (2002a)
Serratia sp. strain HI10	Soil	Matsumura et al. (2009)
Sphingomonas sp. strain BP-7	Seawater	Sakai et al. (2007)
Sphingomonas sp. strain MV1	Wastewater treatment system	Lobos et al. (1992)
Sphingomonas sp. strain TTNP3	Wastewater treatment system	Tanghe et al. (1999)
Sphingomonas bisphenolicum strain AO1	Soil	Sasaki et al. (2005)
Streptomyces sp.	Water	Kang et al. (2004)

within 6 h (Sasaki et al. 2005). Some efficient BPA-degraders belong also to *Pseudomonas* sp. *Pseudomonas paucimobilis* strain FJ-4 degraded 100 BPA/L within 12 h (Ike et al. 1995), while *Pseudomonas monteilii* strain N-502 completely degraded 500 mg BPA/L within 10 days (Masuda et al. 2007). BPA at 60 and 120 mg/L was also easily biodegraded by *Aeromonas hydrophilia* within 6 days according to first-order kinetics (Gulnaz and Dincer 2009).

Microorganisms with a high BPA degradation potential are usually found in industrial and human-destroyed environments. *Enterobacter gergoviae* strain BYK-7 was isolated from outlets of petrochemical wastewater and identified as a very promising organism for BPA removal from industrial streams (Badiefar et al. 2015). The investigated strain tolerated up to 2000 mg BPA/L and degraded on average 23.10 ± 0.126 mg BPA/L in a basal medium within 8 h, 31.35 ± 4.05 mg BPA/L in petrochemical wastewater within 72 h and 53.50 ± 0.153 mg BPA/L in nutritious medium within 48 h.

BPA degradation was identified in mainly Gram-negative bacteria, however, Gram-positive bacterial strains classified as Bacilli can also degrade BPA and possess different degradation potential. *Bacillus* sp. NO13 and NO15 were able to completely degrade 115 mg BPA/L while strain YA27 degraded only 50 mg BPA/L (Matsumura et al. 2009). *Bacillus* sp. strain KU3 isolated from seawater and sediment samples collected from coastal regions of India degraded 74 ± 2% out of 1000 ppm BPA in mineral medium after 12 days of incubation under 150 rpm at 37 °C (Kamaraj et al. 2014). A facultative anaerobic strain, *Bacillus* sp. GZB, was isolated and identified to effectively degrade BPA under both anaerobic and aerobic conditions. Under anaerobic conditions, this strain has used Fe(III) as an electron acceptor. Under optimal aerobic conditions, 5 mg BPA/L was fully removed by *Bacillus* sp. GZB and 51% of BPA was mineralized (Li et al. 2012).

To date, information on BPA degradation intermediates is still limited. Metabolism of BPA was analyzed in detail for *Sphingomonas* sp. strain MV1 (Lobos et al. 1992; Spivack et al. 1994). In the first step of degradation, *Sphingomonas* sp. strain MV1 predominantly metabolized BPA to 1,2-bis (4-hydroxyphenyl)-2-propanol, formed by the hydroxylation of the quaternary carbon in BPA molecule (Fig. 4.1). 1,2-bis(4-hydroxyphenyl)-2-propanol was probably dehydrated to 4,4′-dihydroxy-α-methylstilbene, and then the oxidative cleavage took place, leading to the formation of 4-hydroxybenzaldehyde (HBAL) and 4-hydroxyacetophenone (HAP). HBAL was then subsequently oxidized to 4-hydroxybenzoic acid (HBA). Finally, both HAP and HBA were either completely metabolized by *Sphingomonas* sp. strains MV1 to carbon dioxide or built in the biomass. In the minor pathway, BPA through the solvolysis of the first phenonium ion intermediate was catabolized into to 2,2-bis(4-hydroxyphenyl)-l-propanol formed by the hydroxylation of methyl group. Due to oxidation and rearrangement in 2,2-bis(4-hydroxyphenyl)-l-propanol, another intermediate was formed which then has the C_1 hydroxyl positioned for internal displacement to form the oxirane. Hydrolysis of the oxirane mainly led to the production of the 2,3-bis (4-hydroxyphenyl)-1,2-propanediol, which was further degraded very slowly into 4-hydroxybenzoic acid (HBA) and 4-hydroxyphenacyl alcohol. A minor pathway

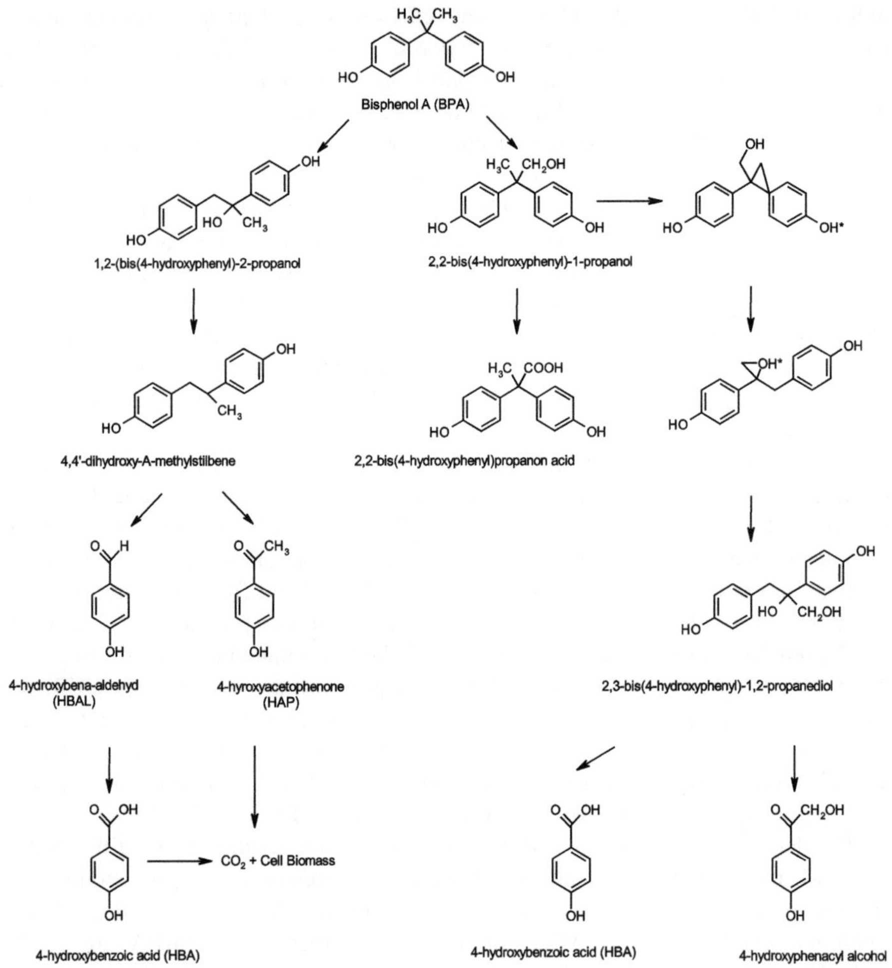

Fig. 4.1 BPA metabolism in *Sphingomonas* sp. strain MV1 (Lobos et al. 1992; Spivack et al. 1994)

to further oxidize the first phenonium ion intermediate was the formation of 2,2-bis (4-hydroxyphenyl)propanoic acid. It was concluded from the balance of total carbon that 60% of the carbon was mineralized to CO_2, while the rest was used in equal amounts for growth of bacterial cell and conversion to soluble organics (Lobos et al. 1992; Spivack et al. 1994). Similar balance of total carbon was also observed for *Bacillus* sp. GZB (Li et al. 2012).

The metabolic process of *Sphingomonas* sp. strain AO1 is similar to that of strain *Sphingomonas* sp. strains MV1 in that 1,2-bis(4-hydroxyphenyl)-2-propanol and 2,2-bis(4-hydroxyphenyl)-1-propanol were also identified as two BPA degradation products (Sasaki et al. 2005). However, degradation of BPA by culture of

Sphingomonas sp. strain BP-7 differs since it resulted in production of 4-hydroxyacetophenone (4-HAP), while 4-hydroxybenzoic acid (4-HBA) was not detected. Slower degradation of 4-HAP than of 4-HBA by strain BP-7 suggests that the degradation pathway of 4-HAP is a rate-limiting step in the degradation of BPA, because of either low activities of initial enzymes in the pathway or inhibition of the growth of the strain by 4-HAP (Sakai et al. 2007). In comparison with strain MV-1, *Sphingomonas* sp. strain BP-7 degraded BPA without accumulation of 4-HAB and degradation of BPA and 4-HAP could have been improved by the addition of nutrients.

C. basilensis strain JF1 and *Sphingomonas* sp. strain TTNP3 mineralize BPA in a similar pathways with 4-(2-hydroxypropan-2-yl)phenol, 4-isopropenylphenol and 4-isopropylphenol as the major intermediates (Kolvenbach et al. 2007; Fischer et al. 2010). The first step of BPA degradation in *Sphingomonas* sp. strain TTNP3 was to introduce one atom of molecular oxygen into the BPA by the NADPH and FAD-aided monooxygenase to form a quinol intermediate (Kolvenbach et al. 2007). Quinol intermediate was mainly further metabolized by *Sphingomonas* sp. strain TTNP3 through the cleavage of the C–C bond between the phenolic moiety and the isopropyl group of BPA to form *p*-hydroquinone (HQ) and a carbocationic isopropylphenol. HQ was further degraded into organic acids. The carbocationic isopropylphenol reacted with water to form 4-(2-hydroxypropan-2-yl)phenol or formed 4-isopropenylphenol and 4-isopropylphenol by the way of the loss of H^+ and the addition of H^-, respectively, as the side reactions. In *C. basilensis*, the ring of BPA is cleaved to 4-(2-propanol)-phenol and HQ. HQ is mineralized while 4-(2-propanol)-phenol is mainly metabolized to 4-isopopenylphenol by the elimination of H_2O at the phenol residue or converted to 4-hydroxyacetophenone. Both these metabolites are then sequentially oxidized to 4-hydroxylbenzaldehyde (HBAL) and then to 4-hydroxybenzoic acid (HBA), which is mineralized (Fischer et al. 2010).

BPA biodegradation in *Bacillus* sp. GZB under aerobic conditions resulted in seven chemicals tentatively identified as the intermediates of BPA biodegradation: 4-(2-hydroxypropan-2-yl)phenol, 4-(prop-1-en-2-yl)phenol, 1-(4-hydroxyphenyl) ethanone, 4-hydroxybenzaldehyde, benzoic acid, 2-hydroxypropanoic acid and 2-methylbutanoic acid (Li et al. 2012). The first four intermediates were verified in many other photocatalytic and biotransformation systems for BPA degradation (Lobos et al. 1992; Ohko et al. 2001; Sasaki et al. 2005; Zhang et al. 2007; Nomiyama et al. 2007; Fischer et al. 2010). It seems that one benzene ring of BPA was first cleaved and hydroxylated to form 4-(2-hydroxypropan-2-yl)phenol and p-benzenediol. 4-(prop-1-en-2-yl)phenol might have either come from dehydration of 4-(2-hydroxypropan-2-yl)phenol or directly from the cleavage of BPA, which was also frequently found in photocatalytic degradation of the BPA pathway (Nomiyama et al. 2007). Similarly, 1-(4-hydroxyphenyl)ethanone might be the daughter product from the oxidation of either 4-(2-hydroxypropan-2-yl)phenol or 4-(prop-1-en-2-yl)phenol. Demethylation of 1-(4-hydroxyphenyl)ethanone produced 4-hydroxybenzaldehyde which was then converted to benzoic acid. Cleavage of benzoic acid resulted in production of small molecular products

2-hydroxypropanoic acid and 2-methylbutanoic acid. The hydroxylation phenomenon might be due to the contribution from the cytochrome P450 monooxygenase during the biodegradation of BPA (Sasaki et al. 2005).

BPA metabolic conversions have also been analyzed in complex, multi-species bacterial communities. When a bacterial consortium that was collected from river sediment was exposed to BPA, following intermediates of BPA degradation were identified after two days of the exposure: 2,2-bis(4-hydroxyphenyl)-1-propanol, 1,2-bis(4-hydroxyphenyl)-2-propanol, carbocationic isopropylphenol, 4-isopropenylphenol, 4,4-dihydroxy-α-methylstilbene and 2,2-bis(4-hydroxyphenyl) propanoic acid (Peng et al. 2015). These intermediates were previously identified as BPA degradation products from *Sphingomonas* sp. strains, *Achromobacter xylosoxidans* and *Cupriavidus basilensis* but not from *Pseudomonas* sp. and *Bacillus* sp. (Telke et al. 2009; Fischer et al. 2010; Li et al. 2012). The presence of BPA degradation products of cytochrome P450 (1,2-bis(4-hydroxyphenyl)-2-propanol and 2,2-bis(4-hydroxyphenyl)-1-propanol) and lack of the degradation products of laccase (2,3-bis(4-hydroxyphenyl)-2-hydroxypropionaldehyde, 2,3-bis(4-hydroxyphenyl)-2,3-dihydroxypropionaldehyde and p-1,2-dihydroxyisopropyl phenol) indicated that the BPA transformation in the analyzed community more likely occurred through the reaction of monooxygenase than laccase. BPA present in river waters has been also metabolized to 2,2-bis(4-hydroxyphenyl)-1-propanol, 2,2-bis(4-hydroxyphenyl) propanoic acid, and 2-(3,4-dihydroxyphenyl)-2-(4-hydroxyphenyl)propane (Suzuki et al. 2004).

Microbial conversions of BPA usually produce metabolites that do not cause or cause lesser (e.g. 4-hydroxyacetephenone) toxic and estrogenic effects than BPA (Ike et al. 2002; Peng et al. 2015). Estrogenicity can temporarily increase and then decline during BPA degradation (Li et al. 2012). However, some intermediates that are more toxic than BPA, such as *p*-hydroxyacetophenone, hydroxybenzaldehyde, *p*-hydroxybenzoic acid, hydroquinone and phenol are known.

There is a variety of evidence indicating that in BPA degradation the cytochrome P450 is involved. Bacterial cytochrome P450 catalyzes hydroxylation, epoxidation, sulfoxidation and dealkylation of a wide range of xenobiotics in cooperation with specific electron transport systems; it degrades drugs, perfumes, carcinogens, pollutants and pesticides. In the culture of *Sphingomonas* sp. strain AO1 culture, addition of cofactors such as NADH or NADPH that support oxidation reactions by cytochrome P450 increased BPA-degradation activity (Sasaki et al. 2005). Inhibition of cytochrome P450 by metyrapone resulted in inhibition of BPA degradation.

Genes *bisdA*, encoding ferredoxin$_{bisd}$ (Fd$_{bisd}$) and *bisdB* encoding cytochrome P450$_{bisd}$ (P450$_{bisd}$) in genome *Sphingomonas bisphenolicum* strain AO1 are heavily involved in BPA breakdown, as shown by molecular experiments (Sasaki et al. 2008). Two transposase genes, *tnpA1* and *tnpA2*, were located upstream and downstream of the region containing *bisdA* and *bisdB* genes. Analysis of amino acid sequence of cytochrome P450$_{bisd}$ has showed that in the cytochrome there are two conserved regions corresponding to the oxygen and heme binding regions. Fd$_{bisd}$ was similar to putidaredoxin-type [2Fe-2S] ferredoxins. P450$_{bisd}$

monooxygenase system, encoded by *bisdAB*, is one system required for BPA hydroxylation in *S. bisphenolicum* strain AO1. Purified components of the strain AO1 cytochrome P450 monooxygenase system, which consists of cytochrome P450 ($P450_{bisd}$), ferredoxin (Fd_{bisd}) and ferredoxin reductase (Red_{bisd}), were involved in initial hydroxylation of BPA and converted BPA to 1,2-bis (4-hydroxyphenyl)-2-propanol and 2,2-bis(4-hydroxyphenyl)-1-propanol, suggesting that the $P450_{bisd}$ monooxygenase system catalyzes the first step of two BPA degradation pathways. Introduction of plasmid pET19b bearing *bisdB* and *bisdAB* genes to *Escherichia coli* BL21 cells enabled BPA degradation by recombinant bacteria. *BisdB*- and *bisdAB*-recombinant cells cultivated in L-BPA medium produced one BPA metabolite—1,2-bis(4-hydroxyphenyl)-2-propanol. Genetic modification of bacteria that are able to metabolize BPA can additionally improve their efficiency of BPA removal—recombinant strain of *Enterobacter gergoviae* [pBR*bisd*] has degraded 45.0 ± 0.3 mg BPA/L in basal medium within 48 h (Badiefar et al. 2015).

The bacterial proteomes of *Sphingobium* sp. BiD32 previously isolated from activated sludge was investigated in the presence and absence of a BPA using label-free quantitative proteomics (Zhou et al. 2015). Out of 1174 of quantified proteins, 184 had a significantly changed abundance in response to the presence/absence of BPA. In the presence of BPA, proteins encoded by genes previously identified to be responsible for protocatechuate degradation were upregulated. BPA degradation by *Sphingobium* sp. BiD32 generated a hydroxylated metabolite and a novel *p*-hydroxybenzoate hydroxylase enzyme was implicated in the metabolic pathway associated with the detected metabolite. Thus, *p*-hydroxybenzoate hydroxylase may be involved in BPA degradation by *Sphingobium* sp. BiD32 and serve as genetic marker for BPA degradation.

The most efficient BPA biodegradation is usually observed in complex microbial communities, in which microorganisms cooperate with each other and the range of metabolic pathways for biodegradation of hard-to-degrade compounds is wide. Synergistic cooperation between bacteria is especially important in nutrient-poor environments such as seawater. Matsumura et al. (2015) have shown that BPA biodegradation in soil resulted from independent activity or mutual cooperation of Sphingomonadales, Xanthomonadales, Burkholderiales and Pseudomonadales. Sphingomonadales, Xanthomonadales and Pseudomonadales were more active in the early stage of cultivation (1–14 days) while the presence of Burkholderiales increased in the middle stage of the process, pointing out to their role in degradation of BPA metabolites. Bioaugmentation of microbial consortium in soil with *Sphingomonas bisphenolicum* AO1 enhanced the efficiency of BPA at concentrations ranged from 1 to 10 mg BPA/g of soil; however, addition of strain AO1 suppressed the growth of Xanthomonadales and Pseudomonadales. At the same time, Burkholderiales and Enterobacteriaceae persisted and probably supported degradation of BPA metabolites or growth of strain AO1. BPA was harmful to microbial communities in soil at concentrations over 10 mg BPA/g of soil. At this concentration, a decrease in microbial diversity was observed.

The microbial community in river sediment amended with different levels of BPA (0, 180 and 450 mg BPA/kg sediment) could nearly deplete all BPA after 3–4 days of incubation (Yang et al. 2014). Gammaproteobacteria and Alphaproteobacteria were the predominant bacterial groups in BPA-degrading sediment microcosm and the higher level of BPA dosage decreased sediment bacterial diversity. The identified bacterial phyla included Proteobacteria, Actinobacteria, Bacteroidetes, Chloroflexi and Firmicutes. *Novosphingobium* was the only known BPA-degrader indentified in the study; however, the authors pointed out to possible involvement of *Sphingobium* sp., *Lysobacter* sp. and *Steroidobacter* sp. in BPA degradation since these microorganisms have been linked to removal of a variety of phenolic compounds.

BPA-degrading activity of certain bacteria can be supported by the presence of other non BPA-degrading species (Sakai et al. 2007). *Pseudomonas* sp. strain BP-14, *Pseudomonas* sp. strain BP-15 and a BPA-degrading *Sphingomonas* sp. strain BP-7 have been isolated from seawater. *Pseudomonas* sp. strain BP-14 did not degrade BPA but shortened the degradation of 100 ppm of BPA by *Sphingomonas* sp. strain BP-7 from 40 to 7 days in SSB-YE medium. Moreover, in a mixed culture of *Pseudomonas* sp. strain BP-14 and *Sphingomonas* sp. strain BP-7, 4-hydroxyacetophenone has not been accumulated.

The adaptation of the microbial community to BPA plays a major role in effective BPA removal. Assessment of the BPA biodegradation ability of various microorganisms from river sediment showed that acclimated microcosms completely degraded 10 mg BPA/L within 28 h and half of the compound within 6 h (Peng et al. 2015). The microbial community changed during degradation and in the final stage *Pseudomonas knackmussii* and *Methylomonas clara* predominated. The degradation ability of mixed strains isolated from the river sediment was higher than that of a single strain but far less than that of the microbial consortium. The role of adaptation of a mixed microbial consortium in activated sludge to BPA degradation has also been raised by Ferro Orozco et al. (2016). In the activated sludge community, the rates of biodegradation increased in the following order: BPA $\ll$ 4HAP $<$ 4HB $\leq$ 4HBA. In the mixed microbial consortium, the oxidation of BPA was the rate limiting step; accumulation of its metabolic intermediates was negligible.

Microorganisms involved in biodegradation of BPA include ammonium-oxidizing bacteria (AOB), which possesses ammonia monooxygenase (AMO), an enzyme that oxidizes ammonia to nitrite. AMO has been shown to use a wide range of non-specific substrates such as hydrocarbons or steroidal estrogens (Chang et al. 2003; Shi et al. 2004). The ability of AMO to metabolize BPA was shown in a pure culture of *Nitrosomonas europaea* (Roh et al. 2009). *N. europaea* was able to degrade BPA with over 80% efficiency; however, addition of allylthiourea, which is an effective inhibitor of AMO, inhibited the process, suggesting that AMO is responsible for the degradation of BPA. The same phenomenon was observed in a mixed consortium of nitrifying activated sludge. 1 mg/L of BPA was degraded by activated sludge within 3 days in the presence of AMO inhibitor and 2 days without it. The role of AOB in BPA degradation will be influenced by the overall composition of the microbial community in a wastewater treatment system. In a reactor with mixed consortia of immobilized microorganisms with high nitrification activity, the contribution of AOB to BPA biodegradation was limited; BPA was mainly removed by heterotrophic bacteria. Exposure of biomass to BPA influenced both the quantity and diversity of AOB in the immobilized biomass (Zielińska et al. 2014). The transformation of BPA in a batch system with *N. europaea* may also proceed by abiotic nitration between biogenic nitrite and BPA (Qian et al. 2012). The rate of transformation of BPA into nitro- and dinitro-BPA depended on nitrite concentrations, temperature and pH values. The nitro- and dinitro-BPA that were formed had much less estrogenic activity than their parent compound BPA. The possible intermediates of abiotic nitration of BPA were 4-n-nonylphenol and 4-n-octylphenol.

The literature reports many factors that influence BPA metabolism in bacterial cells. BPA removal can be efficiently accomplished only under aerobic conditions (Kang and Kondo 2002a). BPA biodegradation by bacteria depends on the temperature. Most of strains degrade BPA efficiently at temperature of 35–37 °C. At temperatures higher than 37 °C, bacterial growth is usually inhibited and BPA degradation activity diminishes (Zhang et al. 2007; Li et al. 2012). At 30 and 20 °C, the half-lives for biodegradation of BPA by microorganisms present in surface river waters (2000–10,000 CFU/mL) were about 3 and 5 days, respectively. At 4 °C, only 20% of BPA (0.04 mg/L) was removed after 20 days of incubation (Kang and Kondo 2002b). At pH 7.0, BPA was degraded most effectively by bacterial strains that have a high BPA tolerance and were isolated from seawater; effectiveness decreased at lower pH (Kamaraj et al. 2014). BPA degradation by *Pseudomonas monteilii* was improved by the presence of Ca(II), Mg(II) and folic acid addition (Masuda et al. 2007).

Bacteria that degrade BPA can be biostimulated (El Fantroussi and Agathos 2005). Phenol stimulated the growth of *C. basilensis* JF1, yet BPA was not degraded co-metabolically. This bacterium completely consumed 1.7 mM phenol in the first 2 h and then degraded 90% of 38.8 mg BPA/L after 225 days. Increased BPA degradation was caused by the higher number of cells that had grown on the phenol (Fischer et al. 2010). Many bacteria need nutrient supplementation to effectively degrade BPA. Nutrient supplementation can overcome inhibition of bacterial growth by some metabolites of BPA degradation such as 4-hydroxyacetophenone (Sasaki et al. 2005). *Sphingomonas* sp. strain BP-7 did not degrade BPA on a mineral salt SSB medium but after the addition of nutrients such as peptone, beef extract and yeast extract to the mineral salt medium, the strain regained its BPA-degradation activity (Sakai et al. 2007). In peptone-supplemented medium, *Sphingomonas* sp. strain BP-7 started to grow after 1 day of cultivation, and BPA in the medium was completely degraded in 4 days. The degradation of the intermediate 4-hydroxyacetophenone was also stimulated. The presence mineral salts and ethanol stimulated BPA degradation by *E. gergoviae* (Badiefar et al. 2015). Bacterial strains having high BPA tolerance were isolated from seawater and sediment samples collected from coastal regions of India. *Pseudomonas* sp. strain KU1, *Pseudomonas* sp. strain KU2, and *Bacillus* sp. strain KU3 showed degradation efficiencies of 78 ± 4%, 81 ± 3%, and 74 ± 2%, respectively, at frequency of rotation of 150 rpm (12 days at 37 °C) in mineral medium supplemented with 1000 ppm BPA. The addition of sodium glutamate as a co-substrate increased the degradation efficiency to 85 ± 3% (strain KU2), 83 ± 2% (strain KU1) and 81 ± 3% (KU3) (Kamaraj et al. 2014).

Immobilization can enhance the activity of BPA-degraders. Bai et al. (2013) investigated the biodegradation of BPA by microorganisms immobilized on polyvinyl alcohol microspheres prepared by the inverse suspension crosslinked method. The predominant genera in the microspheres were *Pseudomonas* and *Brevundimonas*. The main biodegradation products of BPA, by both free and immobilized microorganisms, were 4-hydroxyacetophenone, 2,2-bis(4-hydroxyphenyl)-1-propanol and 2,3-bis (4-hydroxyphenyl)-1,2-propanediol. The BPA degradation studies were carried out at initial BPA concentrations ranging from 25 to 150 mg/L. The affinity constant K_s and maximum degradation rate $R_{\max}$ were 98.3 mg/L and 19.7 mg/(mg VSS d) for free, and 87.2 mg/L and 21.1 mg/(mg VSS d) for immobilized microorganisms, respectively. Higher values of R_{max}/K_s by immobilized microorganisms point out to higher substrate degradation ability of immobilized cells.

4.2 BPA Degradation by Fungi

One of the potential alternatives to effectively degrade BPA is the use of the oxidative action of extracellular fungal enzymes. Many studies indicate fungi with an ability to degrade BPA in soils and aqueous environments (Table 4.2). Similarly to bacteria, there are only a few species of fungi that can effectively metabolize BPA. Chai et al. (2005) studied BPA biodegradation by 26 fungi using an initial BPA concentration of 40 ppm in an aqueous solution in the dark for 14 d. Among the 26 strains tested, 4 did not grow on BPA at all, 11 degraded BPA at 50% or more while *Fusarium sporotrichioides* NFRI-1012, *F. moniliforme* 2-2, *Aspergillus terreus* MT-13 and *Emericella nidulans* MT-98 degraded BPA with more than 99% efficiency. Some strains of white rot basidiomycetes fungi are able to remove 100% of BPA-associated estrogenic activity within 2 h (Cabana et al. 2007). *Aspergillus* sp. isolated from tannery WWTP effluent was able to metabolize BPA as a carbon source for its growth in a concentration range from 20 to 100 ppm (Kamaraj et al. 2012). At 20 ppm of BPA and 9.0 pH out of 3–11 range, *Aspergillus* sp. growth was optimal.

The ligninolytic enzymes are regarded as main enzymes involved in BPA biodegradation that enable removal of its estrogenic activity (Tsutsumi et al. 2001; Suzuki et al. 2003). Due to the complex structure of lignin, its biodegradation system is considered highly nonspecific and ligninolytic enzymes can degrade structurally different environmental pollutants. The ligninolytic enzymes comprise lignin peroxidase (LiP), manganese peroxidase (MnP) and laccase. One of most studied white rot fungi, which mainly produce manganese peroxidase (MnP) and

Table 4.2 Fungi capable of BPA biodegradation

Strain or species	References
Aspergillus fumigatus	Yim et al. (2003)
Aspergillus terreus MT-13	Chai et al. (2005)
Aspergillus oryzae NFRI-1571	Chai et al. (2005)
Aspergillus sydowii KF-17	Chai et al. (2005)
Aspergillus ustus MT-3	Chai et al. (2005)
Byssochlamys fuluva NFRI-1226	Chai et al. (2005)
Curvularia lunata FUTA-N	Chai et al. (2005)
Emericella nidulans MT-98	Chai et al. (2005)
Fusarium graminearum NFRI-1280	Chai et al. (2005)
Fusarium moniliforme 2-2	Chai et al. (2005)
Fusarium sporotrichioides NFRI-1012	Chai et al. (2005)
Heterobasidium insulare	Lee et al. (2005)
Paecilomyces lilacinus IFO-31847	Chai et al. (2005)
Penicillium citrinum NFRI-1019	Chai et al. (2005)
Penicillium expansum NFRI-1021	Chai et al. (2005)
Penicillium frequentans NFRI-1022	Chai et al. (2005)
Phanerochaete chrysosporum ME-446	Tsutsumi et al. (2001)
Pleurotus ostreatus O-48	Hirano et al. (2000)
Rhizopus stoonifer NFRI-1030	Chai et al. (2005)
Stereum hirsutum	Lee et al. (2005)
Trametes versicolor IFO-6482	Suzuki et al. (2003)
Trametes versicolor IFO-7043	Tsutsumi et al. (2001)
Trametes villosa	Fukuda et al. (2001)
Trichoderma viride MT-40	Chai et al. (2005)

lignin peroxidase (LiP) is *Phanerochaete chrysosporium* (Mishra and Pandey Lata 2007). MnP is a heme peroxidase that oxidizes phenolic compounds in the presence of Mn(II) and H_2O_2. This enzyme metabolizes BPA to phenol, 4-isopropenylphenol, 4-isopropylphenol and hexestrol (Hirano et al. 2000). Laccase is a multi-copper oxidase and catalyzes one-electron oxidation of phenolic compounds by reducing oxygen to water. BPA metabolism by laccase includes the polymerization of BPA for forming oligomers, followed by either the addition of phenol moieties or the degradation of the oligomers to release 4-isopropenylphenol (Uchida et al. 2001). A BPA dimmer with a high molecular weight 5,5′-bis-[1-(4-hydroxyphenyl)-1-methyl-ethyl]-bisphenyl-2,2′-diol has been also identified. Degradation products of laccase are 2,3-bis(4-hydroxyphenyl)-2-hydroxy-propionaldehyde, 2,3-bis(4-hydroxyphenyl)-2,3-dihydroxypropionaldehyde and p-1,2-dihydroxyisopropyl phenol. Laccases in different fungi species possess a different ability to degrade BPA.

Peroxidases are haem-containing enzymes that use hydrogen peroxide as the electron acceptor to catalyse a number of oxidative reactions. Analysis of mechanisms for transformation and removal of BPA from the aqueous phase via oxidative coupling mediated by horseradish peroxidase (HRP) showed that BPA can be effectively transformed into precipitable solid products in HRP-mediated oxidative coupling reactions (Huang and Weber 2005). A total of 13 reaction intermediates and products were identified. It is postulated that two BPA radicals are coupled primarily by the interaction of an oxygen atom on one radical and propyl-substituted aromatic carbon atom on another, followed by elimination of an isopropylphenol carbon cation. All intermediates or products detected can be interpreted as resulting from either coupling or substitution reactions between BPA and other intermediates or products. The efficiency of BPA elimination by peroxidase is positively correlated with the level of H_2O_2 in the environment. Peroxidase from *Coprinus cinereus* efficiently removed BPA in aqueous solutions at a molar ratio of H_2O_2 to BPA of about 2.0, pH 9–10 and at 40 °C (Sakurai et al. 2001). The amount of peroxidase needed to efficiently remove 100 mg BPA/L increased with decreasing temperature from 3 U/mL at pH 10 and 40 °C to 5 U/mL at pH 7 and 25 °C.

Possible predominant products in a pathway of BPA degradation by a mixture of ligninolytic enzymes are 2,2-methylenediphenol or bis(4-hydroxyphenyl)methane (bisphenol F) or *p*-(benzyloxy)phenol (Gassara et al. 2013). According to the potential products, the proposed pathway will lead to complete BPA mineralization to water and carbon dioxide (Fig. 4.2).

The ability of ligninolytic enzymes (manganese peroxidase, lignin peroxidase and laccase) to degrade BPA can be enhanced by encapsulation. In the experiment of Gassara et al. (2013), ligninolytic enzymes produced by fermentation of apple pomace waste by *P. chrysosporium* were extracted from fermented solid wastes and encapsulated, using various polymers in order to increase the stability of these enzymes. These enzymes were further used to treat the water contaminated with 10 ppm of BPA. The efficiency of degradation of BPA by the free enzymes was 26% while the use of ligninolytic enzymes encapsulated on polyacrylamide hydrogel and pectin ensured 90% efficiency of BPA degradation after 8 h of incubation. The presence of pectin in the formulation significantly enhanced the activity of enzymes while encapsulation protected the enzymes from non-competitive inhibition.

4.3 BPA Degradation by Algae

The third important group of microorganisms involved in BPA degradation is algae. Under light conditions, algae produce reactive oxygen species such as hydroxyl radicals that cause photodegradation of BPA. The efficiency of BPA removal by algae can be very high. *Chlorella vulgaris* was able to metabolize 120 mg BPA/L within 7 days but concentrations of BPA above 20 mg/L inhibited its cell growth

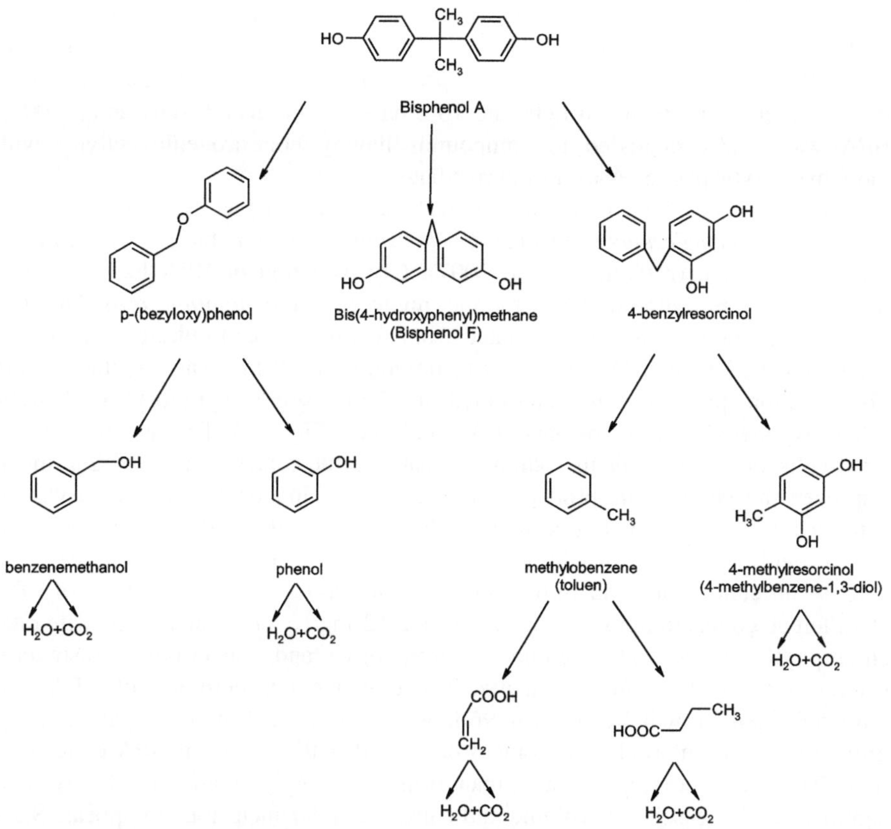

Fig. 4.2 Possible pathways of BPA degradation by fungi

(Gulnaz and Dincer 2009). Representatives of algae species able to degrade BPA are summarized in Table 4.3.

The ability to remove BPA by algae depended on illumination intensity. *Chlorella fusca* was able to remove almost all BPA from landfill leachates in the

Table 4.3 Algae capable of BPA biodegradation

Strain or species	Source	References
Chaetocero gracilis	Institute	Ishihara and Nakajima (2003)
Chlorella fusca	Seawater	Hirooka et al. (2005)
Chlorella sorokiniana	Freshwater	Eio et al. (2015)
Chlorella vulgaris	Freshwater	Gulnaz and Dincer (2009)
Chlorella pyrenoidosa	Freshwater	Zhang et al. (2014)
Chlamydomonas mexicana	Freshwater	Ji et al. (2014)
Nannochloropsis sp.	Institute	Ishihara and Nakajima (2003)
Stephanodiscus hantzschii	Freshwater	Li et al. (2009)

concentration range from 10 to 80 μM for 168 h under continuous illumination at 18 W/m^2. At the light intensity of 2 W/m^2, 82% of 40 μM BPA was removed, and only 27% was removed in the dark. 40 μM of BPA was also efficiently removed (90%) under intermittent 8 h light and 16 h dark conditions (Hirooka et al. 2005). BPA was finally degraded to compounds having nonestrogenic activity with monohydroxybisphenol A as an intermediate.

Recently, it has been reported that freshwater microalgae *Pseudokirchneriella subcapitata*, *Scenedesmus acutus* and *Coelastrum reticulatum* have converted BPA into its glucoside (Nakajima et al. 2007). Glycosylation of BPA has been also investigated using immobilized marine microalga *Amphidinium crassum*. The biotransformation product was isolated from *A. crassum* cell cultures, which had been incubated with BPA for 5 days (Shimoda et al. 2011). The structure of the glycosylation product was determined as 2-(4-β-D-glucopyranosyloxyphenyl)-2-hydroxyphenylpropane (bisphenol A glucoside) (Fig. 4.3). The use of immobilized cells of *A. crassum* in sodium alginate gel at concentrations of 2% much improved the yield of the products. After 5 days of incubation, the glycosylation activity for BPA was increased and the bisphenol A glucoside was produced in higher yield in comparison with the case of the biotransformation using free cells.

BPA influences metabolism of algae. Li et al. (2008) studied the effects of BPA at different concentrations (4, 6, 8, 10 and 12 mg/L) on biomass, growth rate, chlorophyll *a* content, cellular morphology and superoxide dismutase (SOD) activity of *Cyclotella caspia*. The effective concentration of BPA that inhibited algal growth by 50% at 96 h was 7.96 ± 0.23 mg/L. Algal biomass, growth rate and chlorophyll *a* content decreased with increasing BPA concentration. BPA concentrations greater than 6 mg/L strongly inhibited cell division, significantly increased cell volume and caused cellular inclusions to appear. SOD activity increased with the BPA concentration but decreased with prolonged exposure time, showing a dynamic process from induction to inhibition. Growth and chlorophyll *a* accumulation in *Chlamydomonas mexicana* and *Chlorella vulgaris* were inhibited at BPA concentration higher than 10 mg BPA/L, though *C. mexicana* was more tolerant to BPA (a higher EC_{50} value) compared to *C. vulgaris* and more efficiently removed nitrogen and phosphorous compounds (Ji et al. 2014). The highest rates of BPA biodegradation, 24 and 23% by *C. mexicana* and *C. vulgaris*, respectively, were achieved at 1 mg BPA/L after a lag phase of 1 day and 10 day biodegradation time.

Increasing BPA concentration in the environment increases the bioaccumulation of total fatty acid, carbohydrate and BPA in algal cells, revealing that the

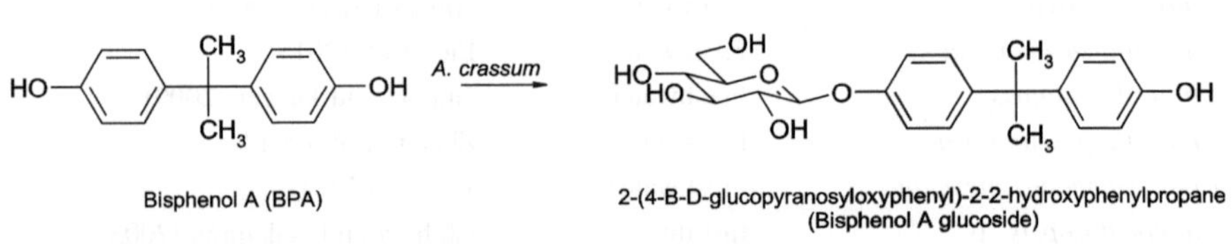

Fig. 4.3 Glycosylation of BPA by cells of *A. crassum* (based on Shimoda et al. 2011)

bioaccumulation of BPA played an important role in its toxicity towards microalgae. BPA tends to accumulate in cellular lipids and induce oxidative stress and lipid peroxydation. BPA accumulation at concentrations in the environment of 25–50 mg BPA/L damages cryo-architecture of cells (Li et al. 2009). Negative influence of BPA on algae metabolism can be reduced in the presence of organics in the environment that may interfere with xenobiotics and modify their effects, modulate algal growth performances or produce a trade-off of both effects. Presence of NOM reduced the BPA toxicity on a green alga *Monoraphidium braunii*, which is known for rapid growth and good tolerance to different natural organic matter qualities. *Monoraphidium braunii* was tested for the ability to tolerate and remove BPA at concentrations of 2, 4 and 10 mg/L, either in NOM-free or NOM-containing media (Gattullo et al. 2012). BPA at the lower concentrations was not toxic for the algae after both 2 and 4 days of exposure, whereas at 10 mg BPA/L it reduced algal growth and photosynthetic efficiency. In the presence of NOM, after 4-day incubation the removal efficiency of 2, 4 and 10 mg BPA/L by *M. braunii* was 39, 48 and 35%, respectively. BPA biodegradation intermediate products were 1-(3-methylbuthyl)-2,3,4,6-tetramethylbenzene and 4-(1-hydroxy-2-methylprop-1-enyl)phenol.

One of the interesting solutions tested to enhance BPA removal efficiency is combining algal and bacterial activity. It has been observed that bacterial growth in the algal-bacterial system reduces the inhibiting effect of BPA on algae (Eio et al. 2015). Normally the growth of *Chlorella sorokiniana* has been inhibited by 20 mg BPA/L, however, this inhibition has not been observed in the algal-bacterial system. In this system, algae provide photosynthetic oxygen that enabled efficient BPA removal by algal-bacterial system while bacteria promote algal growth by CO_2 generation, releasing growth-promoting substances such as vitamins or regulating the pH (Croft et al. 2005). In *C. sorokiniana*-bacterial system, BPA can be degraded abiotically under light conditions to hydroxyl-BPA (Eio et al. 2015), glycosylated to glycosides by algae (Nakajima et al. 2007) and biodegraded by the BPA-degrading bacterial consortium in various pathways. In algal-bacterial system, BPA was degraded to four estrogenic and five nonestrogenic intermediates; the concentration of the first one was lower than 0.4% (Eio et al. 2015).

References

Badiefar L, Yakhchali B, Rodriguez-Couto S et al (2015) Biodegradation of bisphenol A by the newly-isolated *Enterobacter gergoviae* strain BYK-7 enhanced using genetic manipulation. RSC Adv 5:29563–29572

Bai X, Shi H, Ye Z et al (2013) Degradation of bisphenol A by microorganisms immobilized on polyvinyl alcohol microspheres. Front Environ Sci Eng 7:844–850

Cabana H, Jones JP, Agathos SN (2007) Elimination of endocrine disrupting chemicals using white rot fungi and their lignin modifying enzymes: a review. Eng Life Sci 7:429–456

Chai W, Handa Y, Suzuki M et al (2005) Biodegradation of bisphenol A by fungi. Appl Biochem Biotechnol 120:175–182

Chang SW, Hyman MR, Williamson KJ (2003) Cooxidation of naphthalene and other polycyclic aromatic hydrocarbons by the nitrifying bacterium, *Nitrosomonas europaea*. Biodegradation 13:373–381

Croft MT, Lawrence AD, Raux-Deery E et al (2005) Algae acquire vitamin B-12 through a symbiotic relationship with bacteria. Nature 438:90–93

Eio EJ, Kawai M, Niwa C et al (2015) Biodegradation of bisphenol A by an algal-bacterial system. Environ Sci Pollut Res Int 22:15145–15153

El Fantroussi S, Agathos SN (2005) Is bioaugmentation a feasible strategy for pollutant removal and site remediation? Curr Opin Microbiol 8:268–275

Ferro Orozco AM, Contreras EM, Zaritzky NE (2016) Biodegradation of bisphenol A and its metabolic intermediates by activated sludge: stoichiometry and kinetics analysis. Int Biodeter Biodegr 106:1–9

Fischer J, Kappelmeyer U, Kastner M et al (2010) The degradation of bisphenol A by the newly isolated bacterium *Cupriavidus basilensis* JF1 can be enhanced by biostimulation with phenol. Int Biodeter Biodegr 64:324–330

Fukuda T, Uchida H, Takashima Y et al (2001) Degradation of bisphenol A by purified laccase from *Trametes villosa*. Biochem Biophys Res Commun 284:704–706

Gassara F, Brar SK, Verma M et al (2013) Bisphenol A degradation in water by ligninolytic enzymes. Chemosphere 92:1356–1360

Gattullo CE, Bährs H, Steinberg CEW et al (2012) Removal of bisphenol A by the freshwater green alga *Monoraphidium braunii* and the role of natural organic matter. Sci Total Environ 416:501–506

Gulnaz O, Dincer S (2009) Biodegradation of bisphenol A by *Chlorella vulgaris* and *Aeromonas hydrophilia*. J Appl Biol Sci 3:79–84

Hirano T, Honda Y, Watanabe T et al (2000) Degradation of bisphenol A by the lignin-degrading enzyme, manganese peroxidase, produced by the white-rot basidiomycete. Biosci Biotechnol Biochem 64:1958–1962

Hirooka T, Nagase H, Uchida K et al (2005) Biodegradation of bisphenol A and disappearance of its estrogenic activity by the green alga *Chlorella fusca* var. *vacuolata*. Environ Toxicol Chem 24:1896–1901

Huang Q, Weber JW (2005) Transformation and removal of bisphenol A from aqueous phase via peroxidase-mediated oxidative coupling reactions: efficacy, products, and pathways. Environ Sci Technol 39:6029–6036

Ike M, Jin CS, Fujita M (1995) Isolation and characterization of a novel bisphenol A-degrading bacterium *Pseudomonas paucimobilis* strain FJ-4. Jpn J Water Trt Biol 31:203–212

Ike M, Chen MY, Jin CS et al (2002) Acute toxicity, mutagenicity, and estrogenicity of biodegradation products of bisphenol-A. Environ Toxicol 17:457–461

Ishihara K, Nakajima N (2003) Improvement of marine environmental pollution using eco-system: decomposition and recovery of endocrine disrupting chemicals by marine phyto- and zooplanktons. J Mol Catal B 23:419–424

Ji MK, Kabra AN, Choi J et al (2014) Biodegradation of bisphenol A by the freshwater microalgae *Chlamydomonas mexicana* and *Chlorella vulgaris*. Ecol Eng 73:260–269

Kamaraj M, Manjudevi M, Sivaraj R (2012) Degradation of bisphenol A by *Aspergillus* sp. isolated from tannery industry effluent. Int J Phar Life Sci 3:1585–1589

Kamaraj M, Sivaraj R, Venckatesh R (2014) Biodegradation of bisphenol A by the tolerant bacterial species isolated from coastal regions of Chennai, Tamil Nadu, India. Int Biodeter Biodegr 93:216–222

Kang JH, Kondo F (2002a) Bisphenol A degradation by bacteria isolated from river water. Arch Environ Contam Toxicol 43:265–269

Kang JH, Kondo F (2002b) Effects of bacterial counts and temperature on the biodegradation of bisphenol A in river water. Chemosphere 49:493–498

Kang JH, Ri N, Kondo F (2004) *Streptomyces* sp. strain isolated from river water has high bisphenol A degradability. Lett Appl Microbiol 39:178–180

Kolvenbach B, Schlaich N, Raoui Z et al (2007) Degradation pathway of bisphenol A: does *ipso* substitution apply to phenols containing a quaternary α-carbon structure in the *para* position? Appl Environ Microbiol 73:4776–4784

Lee SM, Koo BW, Choi JW et al (2005) Degradation of bisphenol A by white rot fungi, *Stereum hirsutum* and *Heterobasidium insulare*, and reduction of its estrogenic activity. Biol Pharm Bull 28:201–207

Li R, Liu Y, Chen G et al (2008) Physiological responses of the alga *Cyclotella caspia* to bisphenol A exposure. Bot Mar 51:360–369

Li R, Chen GZ, Tam NFY et al (2009) Toxicity of bisphenol A and its bioaccumulation and removal by a marine microalga *Stephanodiscus hantzschii*. Ecotoxicol Environ Saf 72:321–328

Li G, Zu L, Wong PK et al (2012) Biodegradation and detoxification of bisphenol A with one newly-isolated strain *Bacillus* sp. GZB: kinetics, mechanism and estrogenic transition. Bioresour Technol 114:224–230

Lobos JH, Leib TK, Su TM (1992) Biodegradation of bisphenol A and other bisphenols by a gram-negative aerobic bacterium. Appl Environ Microbiol 58:1823–1831

Masuda M, Yamasaki Y, Ueno S et al (2007) Isolation of bisphenol A-tolerant/degrading *Pseudomonas monteilii* strain N-502. Extremophiles 11:355–362

Matsumura Y, Hosokawa C, Sasaki-Mori M et al (2009) Isolation and characterization of novel bisphenol-A-degrading bacteria from soils. Biocontrol Sci 14:161–169

Matsumura Y, Akahira-Moriya A, Sasaki-Mori M (2015) Bioremediation of bisphenol-A polluted soil by *Sphingomonas bisphenolicum* AO1 and the microbial community existing in the soil. Biocontrol Sci 20:35–42

Mishra BK, Pandey Lata AK (2007) Lignocellulolytic enzyme production from submerged fermentation of paddy straw. Indian J Microbiol 47:176–179

Nakajima N, Teramoto T, Kasai F et al (2007) Glycosylation of bisphenol A by freshwater microalgae. Chemosphere 69:934–941

Nomiyama K, Tanizaki T, Koga T et al (2007) Oxidative degradation of BPA using TiO_2 in water, and transition of estrogenic activity in the degradation pathways. Arch Environ Contam Toxicol 52:8–15

Ohko Y, Ando I, Niwa C et al (2001) Degradation of bisphenol A in water by TiO_2 photocatalyst. Environ Sci Technol 35:2365–2368

Peng YH, Chen YJ, Chang YJ et al (2015) Biodegradation of bisphenol A with diverse microorganisms from river sediment. J Hazard Mater 286:285–290

Qian S, Yan L, Pei-Hsin C, Po-Yi P et al (2012) Transformation of bisphenol A and alkylphenols by ammonia-oxidizing bacteria through nitration. Environ Sci Technol 46:4442–4448

Roh H, Subramanya N, Zhao F et al (2009) Biodegradation potential of wastewater micropollutants by ammonia-oxidizing bacteria. Chemosphere 77:1084–1089

Sakai K, Yamanaka H, Moriyoshi K et al (2007) Biodegradation of bisphenol A and related compounds by *Sphingomonas* sp. strain BP-7 isolated from seawater. Biosci Biotechnol Biochem 71:51–57

Sakurai A, Toyoda S, Masuda M et al (2001) Removal of bisphenol A by peroxidase-catalyzed reaction using culture broth of *Coprinus cinereus*. J Chem Eng Jpn 37(2):137–142

Sasaki M, Akahira A, Oshiman K et al (2005) Purification of cytochrome P450 and ferredoxin, involved in bisphenol A degradation, from *Sphingomonas* sp. strain AO1. Appl Environ Microbiol 71:8024–8030

Sasaki M, Tsuchido T, Matsumura Y (2008) Molecular cloning and characterization of cytochrome P450 and ferredoxin genes involved in bisphenol A degradation in *Sphingomonas bisphenolicum* strain AO1. J Appl Microbiol 105:1158–1169

Shi J, Fujisawa S, Nakai S et al (2004) Biodegradation of natural and synthetic estrogens by nitrifying activated sludge and ammonia-oxidizing bacterium *Nitrosomonas europaea*. Water Res 38:2323–2330

Shimoda K, Yamamoto R, Hamada H (2011) Bioremediation of bisphenol A by glycosylation with immobilized marine microalga *Amphidinium crassum*—bioremediation of bisphenol A by immobilized cells. Adv Chem Eng Sci 1:90–95

Spivack J, Leib TK, Lobos JH (1994) Novel pathway for bacterial metabolism of bisphenol A. Rearrangements and stilbene cleavage in bisphenol A metabolism. J Biol Chem 269:7323–7329

Suzuki K, Hirai H, Murata H et al (2003) Removal of estrogenic activities of 17β-estradiol and ethinylestradiol by ligninolytic enzymes from white rot fungi. Water Res 37:1972–1975

Suzuki T, Nakagawa Y, Takano I et al (2004) Environmental fate of bisphenol A and its biological metabolites in river water and their xeno-estrogenic activity. Environ Sci Technol 38:2389–2396

Tanghe T, Dhooge W, Verstraete W (1999) Isolation of bacterial strain able to degrade branched nonylphenol. Appl Environ Microbiol 65:746–751

Telke AA, Kalyani DC, Jadhav UU et al (2009) Purification and characterization of an extracellular laccase from a *Pseudomonas* sp. LBC1 and its application for the removal of bisphenol A. J Mol Catal B Enzym 61:252–260

Toyama T, Sato Y, Inoue D et al (2009) Biodegradation of bisphenol A and bisphenol F in the rhizosphere sediment of *Phragmites australis*. J Biosci Bioeng 108:147–150

Tsutsumi Y, Haneda T, Nishida T (2001) Removal of estrogenic activities of bisphenol A and nonylphenol by oxidative enzymes from lignin-degrading basidiomycetes. Chemosphere 42:271–276

Uchida H, Fukuda T, Miyamoto H et al (2001) Polymerization of bisphenol A by purified laccase from *Trametes villosa*. Biochem Biophys Res Commun 287:355–358

Yang Y, Wang Z, Xie S (2014) Aerobic biodegradation of bisphenol A in river sediment and associated bacterial community change. Sci Total Environ 470–471:1184–1188

Yim SH, Kim HJ, Lee IS (2003) Microbial metabolism of the environmental estrogen bisphenol A. Arch Pharm Res 10:805–808

Zhang C, Zeng G, Yuan L et al (2007) Aerobic degradation of bisphenol A by *Achromobacter xylosoxidans* strain B-16 isolated from compost leachate of municipal solid waste. Chemosphere 68:181–190

Zhang W, Xiong B, Sun WF et al (2014) Acute and chronic toxic effects of bisphenol A on *Chlorella pyrenoidosa* and *Scenedesmus obliquus*. Environ Toxicol 29(6):714–722

Zhou NA, Kjeldal H, Gough HL et al (2015) Identification of putative genes involved in bisphenol a degradation using differential protein abundance analysis of *Sphingobium* sp. BiD32. Environ Sci Technol 49:12232–12241

Zielińska M, Cydzik-Kwiatkowska A, Bernat K et al (2014) Removal of bisphenol A (BPA) in a nitrifying system with immobilized biomass. Bioresour Technol 171:305–313

Chapter 5
Biological Wastewater Treatment Technologies for BPA Removal

Biological degradation is considered one of the most effective ways to reduce the estrogenic activity of BPA in wastewater and to remove BPA from the environment. Indeed, using the metabolic potential of microorganisms to mineralize pollutants is considered a safer and more economic alternative to widely used physico-chemical processes. Although BPA is not readily biodegradable and is considered a pseudo-persistent in the environment, it could be over 90% eliminated using adapted microorganisms. BPA concentrations in the effluents from full-scale wastewater treatment plants (WWTPs) indicate that this compound is not completely degraded during wastewater treatment. Thus, the development of efficient biological methods is crucial.

Although there has been much research on BPA removal in biological wastewater treatment systems, there is still debate about whether biodegradation or biosorption is the most important removal process. In secondary treatment, BPA is biodegraded into less harmful intermediates, or into carbon dioxide and water. However, the relative hydrophobicity of BPA means that it may adsorb onto the surface of sludge. As a result, the sludge may become more estrogenic than the wastewater, which should be taken into account when disposing of excess sludge.

Both biodegradation and biosorption of BPA are described in the context of technologies using activated sludge, biofilm or granular biomass. The efficiencies of various technological systems in WWTPs are presented, as is the fate of BPA adsorbed onto sewage sludge and the risk connected with disposal of this sludge. Finally, methods to detoxify sludge by reusing it to generate value-added products are discussed.

M. ZIELIŃSKA et al., *Bisphenol A Removal from Water and Wastewater*,
https://doi.org/10.1007/978-3-319-92361-1_5

5.1 Removal of BPA by Biomass in Wastewater Treatment Systems

It has been reported that BPA was easily degraded under aerobic conditions (Aguayo et al. 2004), hence there is a lot of data telling about the possibility of BPA biodegradation by activated sludge. Based on the oxygen uptake rate by activated sludge, time necessary for 10% degradation of BPA at a temperature of 22 °C was determined from 4.7 to 5.2 days. After 10 days, between 77.1 and 92.3% of BPA was biodegraded and maximally 93.1% after 28 days (West et al. 2001). At influent BPA concentration of 0.05–550,000 μg/L, Klečka et al. (2001) noticed biodegradation after a short lag-phase of 3.4 days. Microorganisms that are able to biodegrade BPA or utilize BPA as a growth substrate have been isolated and described (Kang and Kondo 2002b). The studies on gram-negative bacteria, strain MV1, growing with BPA as a sole source of carbon and energy in water environment, led to characterization of BPA metabolic pathways (Spivack et al. 1994). Two primary metabolites of BPA biodegradation have been found: 4-hydroxybenzoic acid and 4-hydroxyacetophenone in the major pathway, and 2,2-bis(4-hydroxyphenyl)-1-propanol and 2,3-bis(4-hydroxyphenyl)-1,2-propanediol in the minor pathway (Fig. 4.1). It has been suggested that BPA metabolites are mainly formed through oxidative rearrangement by aerobic bacteria. Many of those metabolites were observed to have similar estrogenicity to BPA (Suzuki et al. 2004), however, the use of *Vibrio fischeri* revealed that biodegradation of BPA is a simultaneous detoxification process (Stasinakis et al. 2008a).

When activated sludge is in contact with xenobiotics, such as BPA, two types of microorganisms, degraders and non-degraders of these xenobiotics are usually found. Even though degraders initially may be absent, the non-degraders can gain the capability of conducting new metabolic pathways through a process called acclimation. The other fraction of non-degraders disappears from the system due to the endogenous decay process. During acclimation, a period of dormant time (lag phase) is followed by a degradation phase of the compound (Chong and Lin 2007).

The type of biomass engaged in wastewater treatment may decide about the efficiency of BPA removal. As compared to activated sludge, systems with immobilized biomass are operated at higher sludge ages and are less sensitive to toxic chemicals. These properties support micropollutant removal from wastewater. During investigations of pharmaceutical residuals from wastewater, Falas et al. (2012) obtained higher rates of micropollutant removal in biofilm than in activated sludge. This was a result of gradients in substrate concentration and redox potential occurring in biofilm that caused stratification of this biofilm. Microorganisms that decompose easily degradable substrates inhabit the outer layers, whilst those that decompose hardly biodegradable substrates inhabit the inner layers of the biofilm. The high concentration and long age of biomass support the growth and activity of slowly growing microbial consortia that break down compounds that are difficult to degrade. In addition, BPA has been reported to be toxic to immobilized biomass at concentrations above 5 mg/L, which is 10 times higher than concentrations that are toxic to activated sludge (Seyhi et al. 2012). The limited diffusion in biofilm layers and the pores of the support reduces the effect of toxic substances on the growth of microorganisms (Zhao et al. 2009). This mechanism does not occur in activated sludge (De Wever et al. 2007) and explains the high toxicity threshold in the biofilm as compared to activated sludge.

There is little data on the removal of BPA in wastewater treatment systems that use aerobic granules. Balest et al. (2008) obtained 93% removal of BPA in a sequencing batch biofilter granular reactor, compared to 72% in the conventional activated sludge process. The reasons for higher BPA removal by granules were high biomass concentration up to 40 g/L, sludge age up to 6 months and low sludge production.

In addition to bacterial cultures, microalgae cultures can be used to remove EDCs. BPA was removed by accumulation in microalgae cells after incubation with a microalgae culture (Nakajima et al. 2007).

5.1.1 Co-metabolism of BPA

Based on Monod-type kinetics, if a compound is present only in trace levels, its removal does not result in any significant biomass growth. These compounds are transformed by co-metabolism, and mixed-substrate growth of the biomass is observed (Schwarzenbach et al. 2003). Co-metabolism is defined as degradation of a compound by microorganisms in which the microorganisms do not receive energy or carbon for cell growth, and thus require an alternative source of carbon and energy (Alvarez-Cohen and McCarty 1991). This process is widespread in nature and causes rapid degradation of compounds that otherwise would degrade very slowly or not at all. The oxidation of these compounds is catalyzed by non-specific enzymes (Speitel and Segar 1995), such as methane monooxygenase and ammonium monooxygenase (AMO). The second enzyme is synthesized by nitrifying bacteria and creates $OH^{\cdot}$ radicals (Wood 1990), which break down resistant organic compounds like BPA. Thus, the activity of nitrifying bacteria, which is stimulated by specific substrates like NH_3, initiates indirect biodegradation of many organic compounds with low molecular weight (Verstraete and Philips 1998). The list of compounds that can be decomposed by ammonium monooxygenase is continuously being updated (Miserez et al. 1999).

Co-metabolism of BPA is likely to take place in bioreactors; this is indicated by the relationship between long solids retention time (SRT), increased nitrifying activity, and simultaneous more effective BPA removal (Clara et al. 2005a). However, it is still unclear whether nitrifying activity is necessary for the biodegradation of BPA. Co-metabolic oxidation by the AMO is probably initiating the biotransformation of endocrine disrupting compounds (EDCs) (Fernandez-Fontaina et al. 2012). There is some evidence for the contribution of nitrifying activity to the biodegradation of BPA. When an ammonium monooxygenase inhibitor (allylthiourea, ATU) was added to a reactor containing *N.*

europaea, they did not degrade BPA (Roh et al. 2009). Furthermore, BPA concentration in the effluent decreases as ammonium is being oxidized to nitrate (Kim et al. 2007). If the ammonium in the influent is replaced with nitrite, an acclimation period is needed before significant degradation of BPA takes place, during which the heterotrophs adapt to using BPA as an energy source. In conventional and advanced treatment in WWTPs, when nitrification efficiency was lower than 90%, median BPA removal was 44%, while median BPA removal was higher than 78% when nitrification efficiency exceeded 90%, indicating the strong influence of nitrifying conditions on BPA removals (Guerra et al. 2015). Ren et al. (2007) observed that with the increase in the COD loading, the rate of transformation of organic matter increased but the efficiency of ammonia oxidation and EDC degradation decreased. This indicated the prevalence of co-metabolism of ammonia-oxidizing bacteria (AOB) in the biodegradation of EDCs in nitrifying activated sludge.

In contrast, it has been reported that co-metabolism was not involved in biodegradation of micropollutants, including BPA (De Wever et al. 2007). Direct biodegradation rather than co-metabolic transformations are promoted by the lower organic loading in membrane bioreactors than in conventional activated sludge, which limits the availability of substrate. The same conclusions were drawn by Stasinakis et al. (2008a), who conducted tests of BPA biodegradation. Their experiments showed the absence of co-metabolic phenomena during BPA biodegradation in the co-presence of a readily biodegradable compound. Biodegradation reached 10% by day 1 and exceeded 60% by day 9, indicating that a prolonged time was required for adequate biodegradation of the mixture, possibly due to incomplete predominance of BPA-degrading bacteria. When BPA was the sole carbon source, it was biodegraded at an efficiency of 87.8 ± 6.9% after an initial lag phase. Degradation reached 10% by day 4.5 and exceeded 60% by day 6.2. Because these results meet the strict definition of ready biodegradability, BPA is not expected to persist in an activated sludge system.

The potential for BPA removal by mixed consortia of immobilized microorganisms of high nitrification activity was investigated with BPA concentrations from 2.5 to 10.0 mg/L at a very short hydraulic retention time (HRT) of 1.5 h (Zielińska et al. 2014). The presence of BPA in the influent limited ammonium oxidation; nitrification efficiency decreased from 91.2 ± 1.3% in the reactor without BPA to 47.4 ± 9.4% when BPA concentration in the influent was highest. A strong negative linear relationship between the efficiency of BPA removal and nitrification efficiency indicated that co-metabolism could not have been responsible for the removal of BPA. Furthermore, oxygen uptake rate measurements were conducted with the presence or absence of ATU that is an inhibitor of ammonia monooxygenase. After ATU was administered, ammonia was not removed but BPA was. The degradation capacity of BPA was comparable independently of the presence of ATU. This implied that BPA removal was due to the activity of heterotrophs. Similar results were obtained by Kim et al. (2007), however, when authors used Hg_2SO_4 as an inhibitor, a decrease in BPA concentration was smaller because of the inactivation of all the biological activity of cells. A prolonged exposure to BPA at concentrations of 2.5–10.0 mg/L or the release of BPA by-products could have caused damage to AOB cells or inactivation of ammonium oxidation (Zielińska et al. 2014). The inhibition of the conversion of ammonium to hydroxylamine by binding to ammonia monooxygenase enzyme as a result of the presence of organic micropollutants was reported in the literature (Alpaslan Kocamemi and Cecen 2007).

Although growth conditions in the biofilm, where oxygen is a factor limiting the biodegradation of pollutants, enhance the ability of bacteria to co-metabolise (Speitel and Segar 1995), other factors may dictate the type of metabolism. The threshold level of micropollutant removal depends on the presence of the minimum level of micropollutant concentration below which the microorganisms cannot gain sufficient energy for growth. This is characteristic for direct biodegradation and not for co-metabolism. The kinetics of co-metabolism are at least 10 times slower than the kinetics of metabolism and may be further decreased by enzyme competition (Speitel and Segar 1995). This competition may occur because metabolism of a growth substrate and co-metabolism of non-growth substrates are catalyzed by the same enzyme. The presence of high amounts of organic substrate means nitrifiers are less efficient in competing for oxygen (Xiangli et al. 2008). Alternative substrates of AMO may inhibit the ammonia oxidation due to their competition for binding to monooxygenase (Roh et al. 2009). This phenomenon has a high impact on the design and operation of bioreactors.

Co-metabolism may occur under starvation conditions. Therefore, to avoid enzyme competition it is very important that microorganisms have the ability to maintain enzyme activity and sufficient energy reserves to support co-metabolism for some period in the absence of the growth substrate. The energy required for co-metabolism may be supplied by storage products, such as poly-β-hydroxybutyrate (Speitel and Segar 1995). These considerations indicate that the AOB role in BPA removal from wastewater may indeed be limited, especially since many known strains of bacteria are capable of degrading BPA (Roh et al. 2009).

Regardless of whether biodegradation of a given compound is direct or indirect, bioreactor performance is dependent on the diversity of particular microbial groups (Daims et al. 2001). The presence of many species of microorganisms, capable of performing specific processes, ensures that wastewater treatment efficiency is maintained, even during a sudden deterioration of environmental conditions. The most resistant species adapt and ensure that specific metabolic pathways remain active (LaPara et al. 2002). The molecular approach in the studies by Zielińska et al. (2014) proved that the increase in influent BPA concentration resulted in the lowering the diversity of AOB in immobilized biomass. This is in agreement with Xia et al. (2008) who stated that an increase in organic loading limits biodiversity because high COD concentrations favor the fast-growing bacteria that predominate over slowly-growing bacteria. So far, however, issues concerning the relationship between diversity and the removal of BPA by technical biota are not widely recognized. At the current stage of research, it is important to broaden knowledge of the mechanisms of BPA removal in biological wastewater treatment systems and the role of specific bacteria, like AOB, in this process. This information may be helpful in the implementation and operation of technologies to increase the efficiency of removal of micropollutants from wastewater.

5.1.2 Sorption of BPA

Although BPA can be degraded by microorganisms, it is hard to completely remove it from wastewater by conventional biological treatment methods. As a result, residual BPA is present in the effluents from municipal wastewater treatment plants in concentrations ranging from 0.01 μg/L (Nasu et al. 2001) to 86.0 μg/L (Kasprzyk-Hordern et al. 2009). One of the reasons that biological treatment is not completely effective is because BPA can be sorbed on suspended solids or biofilm.

The efficiency of micropollutant removal depends on their physical, chemical and biological properties. The values of octanol-water coefficient (log K_{ow}) of 3.32 and activated sludge-water partition coefficient of $K_D > 1000$ L/kg TSS indicate that BPA has a moderate affinity for the solid phase (moderate hydrophobicity), so under some operational conditions, BPA can sorb to biofilm or suspended solids (Staples et al. 1998; Stringfellow and Alvarez-Cohen 1999). Hence, in technological systems, the efficiency of secondary clarifiers decides if the post-treatment to lower the BPA concentration in effluent and limit BPA adverse effects on water ecosystems is necessary.

It has been reported that 15–18% of BPA can be removed as a result of sorption (Clara et al. 2004). Activated sludge sorbed BPA in amount of 0.033–36.7 μg/g TSS (Lee and Peart 2000), 30–330 μg/g TSS (Meesters and Schroeder 2002), and 28.3 μg/g TSS (Shen et al. 2005). Biosorption efficiency depends on the type of bacteria. Filamentous bacteria responsible for sludge bulking had the highest sorption capacity, thus causing the risk of micropollutants being released from the excessive sludge to the environment (Stringfellow and Alvarez-Cohen 1999).

Sorption potential of organic compounds, including BPA, is connected with SRT, HRT and nitrification efficiency. An increase in SRT is expected to increase the partitioning of synthetic organic compounds due to the fact that more hydrophobic and less negatively charged activated sludge flocs can be formed at higher SRT (Langford et al. 2007). SRT from 3 to 20 d did not affect the sorption of BPA (Stasinakis et al. 2010). Although biodegradation was a major BPA removal mechanism in conventional and advanced treatment in WWTPs, sorption was a primary removal pathway under insufficient nitrification conditions (Guerra et al. 2015). When nitrification was increased, biodegradation and sorption were responsible for BPA removal, with biodegradation being predominant. Longer HRTs resulted in a decrease of sorption capacity because degradation was enhanced under these conditions (Guerra et al. 2015).

So far, there have been conflicting reports on the relative contributions of sorption and biodegradation to BPA removal in biological systems. Trace polar organic compounds, like BPA, are sorbed to a limited extent and their elimination is the result of biodegradation, whereas sorption is the main mechanism for the removal of trace non-polar compounds (De Wever et al. 2007). In an activated sludge reactor and a membrane bioreactor, the participation of sorption in BPA removal was low; from 0.004 to 1.35% of introduced BPA was sorbed (Chen et al. 2008). The authors identified 4-hydroxyacetophenone as the primary by-product of BPA oxidation in the liquid phase, which confirmed the dominance of biodegradation in BPA removal. However, sludge sorption might play an important role in BPA removal because it prolongs retention time of micropollutants in the reactor for subsequent biodegradation (Tadkaew et al. 2011). Biodegradation is favored by the long age of biomass and the resulting long time of BPA retention in bioreactors with supports (Zhao et al. 2008). The above findings are in contrast to those of Boonyaroj et al. (2012), who reported that both biodegradation and sorption are important in the removal of BPA from wastewater in the system treating municipal solid waste leachate.

In municipal and industrial wastewater treatment plants, as well as in soil, biodegradation was the main mechanism of the removal of BPA (Pothitou and Voutsa 2008; Samaras et al. 2013; Li et al. 2013). When compared to nonylphenol, 17β-estradiol, 17α-ethinylestradiol, BPA had the lowest potential of sorption onto the soil. Significantly higher efficiency of BPA removal was gained in activated sludge compared to inactivated sludge (Seyhi et al. 2012). In an aerobic WWTP with no BPA accumulation in sludge, 52–100% of BPA removal was observed which showed high BPA solubility with no partitioning to sludge (Fernandez et al. 2009). BPA removal from landfill leachate in a two-stage membrane bioreactor was investigated, achieving almost 100% efficiency after 100 days of the process (Boonyaroj et al. 2012). BPA was mainly detected in a liquid phase, which allowed biodegradation to be considered a main mechanism of BPA removal. Static experiments confirmed this thesis; during 24-h test, 21% of BPA was oxidized and 11% was adsorbed. There is a hypothesis that BPA is biodegraded after sorption. Regardless of the experimental conditions, 58–71% of BPA introduced into the bioreactor was sorbed on immobilized biomass (Zielińska et al. 2014). However, there was no relationship between the amount of BPA sorbed and the efficiency of BPA removal from the liquid phase. This indicated that BPA was initially sorbed on the biofilm and afterwards biodegraded. Increasing the initial BPA concentration from 2.5 to 10.0 mg/L resulted in an increase in the share of BPA in biomass. If sorption had been the main mechanism of BPA removal, the efficiency of BPA removal would have decreased as initial BPA concentration increased because of depletion of the sorption capacity of the solid phase. When investigating BPA adsorption in sludge from an membrane bioreactor in the presence of different BPA concentrations (0.45–18.2 mg BPA/L), Seyhi et al. (2011) observed very low interactions between BPA and sludge, such as Van Der Waals attraction or electrostatic strengths. Membrane filtration coupled with intensive aeration through the membrane (in order to minimize membrane fouling) decreases the floc size. This makes activated sludge from the membrane bioreactor system different from the conventional activated sludge, which affects the pollutant sorption.

BPA was sorbed on the activated sludge flocs the least, when compared with 4-n-nonylphenol and polychlorinated biphenyl (Felis et al. 2011). It has been proven that sorption alone or sorption prior to biodegradation was responsible for the removal of BPA. Due to the different sorption characteristics of activated sludge coming from municipal and industrial WWTPs, not only the affinity of substrate to the sorbent should be determined but also the sorption capacity of particular activated sludge.

Although different biological methods have the potential of BPA removal, the fate of BPA in the treatment process is not evident. The above-mentioned studies show that in conventional systems of wastewater treatment BPA is removed by direct or indirect biodegradation or sorption. Loss of BPA as a result of volatilization in aerated activated sludge tanks is minimal because of low Henry's coefficient for BPA.

5.1.3 Factors Affecting Biological Removal of BPA

The concentration of dissolved oxygen is believed to affect the biodegradation of BPA in wastewater treatment systems. BPA in both suspended and soluble form does not biodegrade well under anaerobic conditions (Ike et al. 2006), whereas its biodegradability increases under aerobic conditions (Kang and Kondo 2002b). Press-Kristensen et al. (2008) noticed that BPA biodegradation occurred only under aerobic conditions. When removing BPA from the effluent of an anaerobic membrane bioreactor, Abargues et al. (2013) noted that aerated experiments produced faster removal than a non-aerated experiment. They found that, under aerated conditions, removal efficiency was more than 91%, whereas under non-aerated conditions, the efficiency was between 50 and 80%. At an initial concentration in the range 0.05–10 μg/L, BPA could be biodegraded anaerobically in 30 days at 9–22 °C, whereas the aerobic biodegradation took only 72 h (Ogawa et al. 2005). In contrast, according to Sarmah and Northcott (2008), fast BPA biodegradation took place both in the presence or absence of oxygen; independently of the conditions, 90% BPA was biodegraded during 2–4 days.

Wastewater temperature was found to affect BPA degradation. 99% BPA removal was observed in summer mainly as a result of biodegradation (Nie et al. 2012). An increase in BPA concentration in the effluent in winter resulted from the lowering of microbial activity and activated sludge concentration at low temperatures. According to Kang and Kondo (2002a), the rate of BPA degradation depends on bacterial counts of the samples, which was connected with water temperature. With the increase in temperature, the bacterial counts and BPA degradation rate increased. During the treatment of rural wastewater by activated sludge, 87.9% of BPA was removed in summer, whereas in winter only 21.3% (Qiang et al. 2013). In summer, the higher removal of BPA could be attributed both to elevated temperatures and longer exposure to sunlight, enhancing biodegradation and photodegradation (Guerra et al. 2015). At influent BPA concentration of 1500 μg/L, sludge age of 27–52 d and aerobic retention time of 0.33–0.5 d, the period of activated sludge adaptation to influent BPA concentration lengthened from 10 days in summer (19–22 °C) to 43 days in winter (15–16 °C) (Press-Kristensen et al. 2008).

The efficiency of BPA removal is connected with SRT, which indicates the mean residence time of microorganisms in the reactor, and HRT, which indicates residence time of wastewater in the bioreactor. Long SRT increases biomass concentration and favors the development of diverse microbial population including slowly-growing microorganisms that have greater capacity of xenobiotic removal (Kim et al. 2005; Guerra et al. 2015). In WWTPs operated at long SRT, biodegradation of the micropollutants and autotrophic ammonia oxidation (nitrification) are observed. Thus, for the design of technological systems treating wastewater containing micropollutants, the SRT is considered the most important parameter. If a micropollutant is biodegradable, in the low concentration and in the case of mixed substrate growth (co-metabolism), a specific SRT can be associated to the substance (Clara et al. 2005a). This specific SRT must be higher than the reciprocal of μ_{tot} (SRT > μ_{tot}^{-1}, μ_{tot}—the total specific growth rate). At SRTs below this critical value, no biodegradation is expected. According to Clara et al. (2005a), high efficiency of BPA removal can be achieved at SRTs longer than 10 days at a temperature of 10 °C. Authors observed no BPA removal at SRT of 1 day and almost 90% removal at SRT of 13 and 26 d. Therefore, it can be concluded that WWTPs with nitrogen removal also efficiently remove degradable micropollutants, such as BPA. In WWTPs operated with activated sludge of low SRT, BPA removal is rather low. At SRT of 4.4 d, partial nitrification was obtained and BPA concentrations in the effluent were higher than at SRT of 7 d, at which complete nitrification was achieved (Drewes et al. 2005). When examining BPA removal in conventional and advanced treatment in WWTPs, SRT showed a better correlation with BPA removal at temperatures <15 °C; no correlation was observed at higher temperatures (Guerra et al. 2015). At temperatures <15 °C, BPA removal was influenced by MLSS concentrations.

The hypothesis that higher SRT enhances BPA removal is not confirmed by Samaras et al. (2013), who investigated BPA removal in two municipal WWTPs using secondary treatment with biological nitrogen and phosphorus removal. The first WWTP was operated with HRT of 9 h, SRT of 8 d, mesophilic anaerobic sludge digestion (SRT 17 d) and thermal drying. The second WWTP was operated with HRT of 23 h, SRT of 18 d and sludge dewatering. Higher BPA removal, due to biodegradation, was achieved in the first WWTP (96%), in which SRT was lower. In the examined range, SRT did not affect BPA biodegradation. Anaerobic digestion in this WWTP could have been responsible for higher total biodegradation. The authors suggest that the research should be more focused on composition and structure of biomass as factors that influence BPA removal more than SRT. According to Stasinakis et al. (2010), BPA removal at a temperature of 24 °C did not depend on SRT in the range of 3–20 d. At influent BPA concentration of 20–30 μg/L, HRT of 10 h and oxygen concentration of 4 mg/L, 89.8–93.8% of BPA was removed by activated sludge. Microorganisms capable of degrading BPA were present in all 3 systems, independently of the SRT. However, at SRT of 3 days, the highest biodegradation rate constants (113 L/(g VSS d)) were observed, while they significantly decreased at SRT of 10 and 20 days (15 and 17 L/(g VSS d), respectively). This indicates that microorganisms capable of degrading BPA were more active in low SRT. Joss et al. (2006) investigated that at SRT of 3 days, the relative amount of heterotrophic biomass per dry sludge matter was almost twice as high as the amount estimated for SRT of 20 days.

A high positive correlation between BPA removal and HRT was shown in conventional and advanced treatment in WWTPs (Guerra et al. 2015). Particularly, the stronger influence of HRTs at lower temperatures was observed, which may be related to a decrease in the activity of microorganisms. The authors suggested that to optimize BPA removal at lower temperatures (<15 °C) a longer HRT may be required; to achieve 80% efficiency, the median HRT should be 21 h.

The efficiency of BPA removal is affected by the concentration of BPA. This efficiency rose from 87.1 ± 5.5% to 92.9 ± 2.9% at BPA concentration in the influent increasing from 2.5 to 10.0 mg/L (Zielińska et al. 2014). This high efficiency of BPA removal could be explained by the fact that its by-products could have been substrates for the growth of other bacteria, which would have created an extensive network of trophic relationships that favors biodegradation (cross-acclimation) (Yuan et al. 2002). A higher concentration of BPA or its metabolites increases the degradation rate and shortens the half-life of BPA (De Wever et al. 2007). In addition, initiation of biodegradation results from the enzymatic adaptation of the microorganisms to the amount of substrate in wastewater (LaPara et al. 2006).

To optimize BPA removal, the food/microorganisms ratio (F/M) ratio should be adjusted. In reactors with high biomass concentration, the F/M ratio is low, which forces the degradation of organic micropollutants. At higher F/M, more easily degradable substrates are metabolized. For this reason, Urase and Kikuta (2005) have postulated that when operating an activated sludge system for removal of micropollutants from wastewater, preliminary elimination of easily degradable organic pollutants should be conducted. The biodegradation of BPA by activated sludge could interfere with the degradation of biodegradable substrates, such as glucose, acetate, peptone and others (Sasaki et al. 2005; Urase and Kikuta 2005; Zhao et al. 2008; Stasinakis et al. 2008a). Thus, the BPA degradation may be diminished by the diauxic growth condition, in which the microbial demand for growth and energy can be obtained from the readily degradable substrate (Chong and Chen 2007). However, some studies showed the beneficial effect of nutrients on BPA degradation. Ferro Orozco et al. (2013) showed that BPA degradation was not affected by the presence of cheese whey or acetate. When acetate was tested, oxygen uptake rate of acclimated activated sludge were about twofold the values of the non-acclimated ones. This enhancement could be attributed to the metabolic pathway of BPA because acetate is an intermediate of the metabolism of BPA (Spivack et al. 1994). The presence of acetate as an intermediate during the metabolism of BPA was believed to cause an acclimation of the activated sludge to BPA.

Biological and chemical processes of the removal of nutrients and easily biodegradable organic compounds may additionally transform trace micropollutants such as BPA. When nitrite is present in wastewater, nitro- and nitroso-phenolic transformation products can be formed under acidic conditions. As a result, these products are emitted via WWTP effluents into the aquatic environment (Jewell et al. 2014). This phenomenon may change the properties of micropollutants; after transformation to dinitro-BPA, the estrogenic activity of BPA decreased while genotoxicity increased (Toyoizumi et al. 2008). At pH of 7, nitrite concentration of 0.6 mg/L and BPA concentration of 1 mg/L, Jewell et al. (2014) observed the formation of NO_2-BPA and dinitro-BPA, probably due to the generation of radicals from nitrous acid. However, under typical conditions at the WWTPs like neutral pH and low nitrite concentration, a significant formation of nitrophenols could be excluded. Under neutral conditions, the transformation of BPA was observed via biotic pathways including hydroxylation and sulfurylation.

In WWTPs that receive a substantial amount of industrial water and stormwater, heavy metals are present. Due to the fact that these metals are an important group of enzymatic activity modulators, they affect enzymatic activity and efficiency of wastewater treatment, thus affecting the fate of BPA during its biodegradation. The presence of Co(II), Zn(II), Ni(II), and Mn(II) did not pose problems for the treatment of real wastewaters with laccase (Kim and Nicell 2006). However, in the presence of Cu(II) and Fe(III), BPA conversion relative to controls was reduced. It may be supposed that there are interactions between heavy metal and BPA and BPA may become more toxic when coming into contact with heavy metal (Mohapatra et al. 2010a). According to Koponen and Kukkonen (2001), BPA can also synergize with cadmium, increasing harm to the organisms. It may combine with heavy metal, form a BPA–metal complex and become adsorbed to the sludge (Pathak et al. 2009).

The acclimation of activated sludge to BPA was examined at SRT of 30 and 45 d using two strategies: constant (at 40 mg BPA/L) or increasing (from 40 to 320 mg BPA/L) the initial BPA concentrations (Ferro Orozco et al. 2013). Both strategies led to acclimation of activated sludge; specific BPA degradation rate of acclimated activated sludge ranged from 65 to 90 mg BPA/(g TSS d). However, the former strategy was more effective than the latter; in the latter, lower specific BPA consumption rates were obtained due to the toxic effect of the high initial BPA concentrations employed.

It has been demonstrated that BPA-degrading bacteria enhanced degradation when added to activated sludge. In the studies by Zhou et al. (2014), addition of *Sphingomonas* sp. cultures in a 4-stage activated sludge process improved BPA degradation by 87.6%. In this context, SRT is a very crucial factor. SRT influences the fate of bioaugmented bacteria that can be longer retained in the system so that lower bioaugmented doses can result in higher reactor concentrations of these bacteria.

The intensification of the biodegradation process is facilitated by the use of immobilized enzymes (biocatalysts). However, in environmental applications, the enzyme is often a cost-determining factor, hence the recycling and recovery of biocatalysts is crucial. In a study by Demarche et al. (2012a), a laccase preparation from *Coriolopsis polyzona* MUCL38443 was immobilized via a sequential adsorption-crosslinking process on mesoporous silica particles. Laccases are oxidoreductases of great interest in environmental biotechnology because they can degrade a broad range of xenobiotics (Demarche et al. 2012b). The biocatalyst was used in a continuous stirred-tank membrane reactor to continuously degrade the BPA in wastewater (Demarche et al. 2012a). A 90% removal of 50 mM BPA was achieved. Laccases catalyze the one-electron oxidation of substrates with the concomitant reduction of molecular oxygen to water. This renders them very attractive compared to other enzyme systems because no additional expensive co-substrate or cofactor is required apart from oxygen (Demarche et al. 2012a). The other advantage of laccase treatment was proved by Cabana et al. (2007) who showed that stable polymers of BPA are formed after a laccase treatment in aqueous media. These water insoluble oligomers of BPA do not exhibit the estrogenic activity of the parent compound and can be easily removed from the liquid phase by, for example, filtration.

5.1.4 BPA Removal in Membrane Bioreactors (MBRs)

The initial concentration of BPA is one of the major factors affecting biodegradation. According to Michaelis-Menten kinetics, some amount of a pollutant is necessary to achieve efficient biodegradation. When a concentration of micropollutant decreases below a certain threshold level, biodegradation may not occur due to lack of enzyme induction (Kovarova-Kovar and Egli 1998). A higher level of BPA and its metabolites promoted biodegradation in activated sludge via accelerated removal speed of BPA and its metabolites, increasing degradation rate and decreasing half-lives of biodegradation (Xie et al. 2011). The degradation of low concentrations of a target pollutant usually required a long incubation time and sometimes had a lag phase. For example, biodegradation half-lives for BPA were 0.5–3 d at the initial concentrations of 50–5500 μg/L, but 3–6 d at environmentally relevant concentrations of 0.05–0.5 μg/L with lag phases of 2–4 days (Klečka et al. 2001). BPA levels in the influent of wastewater treatment plants are usually in the range of ng/L or mg/L, and cannot be degraded effectively in most of wastewater treatment plants (Xie et al. 2011).

To increase the biodegradation efficiency, different technical solutions are required. These include the use of an adsorbent matrix for the accumulation of the pollutant as well as retention of microorganisms in a reactor system (systems with immobilized biomass, membrane bioreactors). Such solutions ensure both availability of higher concentrations of the organic pollutants for biodegradation and maintenance of a microbial population able to bring about the biodegradation. For example, to increase the substrate concentration, Xie et al. (2011) added molecularly imprinted polymeric microspheres (MIPMs) that are selective sorbents to activated sludge. Adsorption of MIPMs to BPA and its analogues increased levels of BPA and its metabolites, which were also the substrates of biodegradation. This enhancement was significant in waters containing trace pollutants, and in waters containing different interferences such as heavy metals and humic acids.

The other solution to improve the removal of organic micropollutants from wastewater is the use of bioreactors with high biomass concentration and long SRT (Boonyaroj et al. 2012). These conditions can be provided in membrane bioreactors (MBRs), in which microorganisms are maintained at much higher concentrations as compared to conventional wastewater treatment systems, due to the complete separation of activated sludge from the effluent. The combined physico-chemical and biological processes in MBRs have yielded much better removal efficiency of micropollutants both with hydrophobic and hydrophilic properties than conventional systems (Boonyaroj et al. 2012). Despite the fact that in the MBRs for wastewater treatment usually micro- and ultrafiltration membranes are used which do not allow retention of BPA due to size exclusion, a tendency of hydrophobic organic substances to accumulation onto the solids makes these membranes barriers for micropollutants.

The possibility of maintaining high retention time of biomass makes the MBRs able to enrich slowly growing bacteria and host diverse microbial biocoenosis with broad physiological capabilities, which can induce degradation (Clara et al. 2005a). For this reason, the MBR technology is suitable for the treatment of wastewater with low BOD/COD ratio (Fatone et al. 2005). A periodical decrease in the efficiency of removal of different micropollutants was observed in the MBR when compared with the conventional activated sludge (De Wever et al. 2007). Despite these variations, a shortened lag-phase was observed in the MBR, which resulted from the high biodiversity. Because of this high biodiversity and high biomass concentration in the MBR, the frequency of interactions between microorganisms is increased and the exchange of genetic information is supported that make adaptation of biomass to environmental conditions faster. To confirm this, micropollutant dosage was interrupted for a month. After this period, microorganisms in the MBR resumed the biodegradation faster than in the activated sludge.

In conventional activated sludge plants, the final efficiency is dependent on the biomass ability for sedimentation in the secondary clarifier. Because of the complete retention of solids, in the MBRs this parameter is of minor significance. Higher biomass concentrations in MBRs results in smaller plant sizes. Disadvantages of the MBRs include higher cost and requirements in operation and maintenance, and higher power consumption compared to conventional activated sludge systems (Clara et al. 2005b).

There are a lot of examples on the promising application of the MBRs in BPA removal from wastewater. A two-stage MBR was used for the treatment of municipal solid waste leachate under no sludge wastage conditions at a HRT of 1 d (Boonnorat et al. 2014). The first anoxic reactor was followed by an aerobic reactor with hollow fiber membrane modules immersed inside the tank for solid-liquid separation. In the anoxic tank, MLSS increased from 3 to 7 g/L, in the aerobic tank it was maintained at 7–9 g/L. In the anoxic tank, 37.7% of BPA was removed, whereas in the whole system 99.5% of BPA was removed without reaching maximum adsorption capacity of sludge. Biodegradation under aerobic conditions was the main mechanism of the removal. In the studies by Nghiem et al. (2009), at influent concentration of 750 μg BPA/L, both biodegradation and adsorption to the sludge were responsible for over 90% removal of BPA. The investigations of BPA removal from landfill leachate in the MBR with UF showed that high removal efficiency was not due to the membrane retention capacity but to biodegradation or association with macro-molecular material (Wintgens et al. 2004).

BPA removal was compared in a conventional activated sludge reactor and MBR, at influent BPA concentrations from 0.1 to 20.0 mg/L (Chen et al. 2008). Although the efficiencies of BPA removal were similar, the MBR could bear much higher volume loadings than a conventional reactor. In MBR, shortening HRT from 11 to 3.9 h did not influence the removal of BPA because a long SRT (350 days in the MBR system), as compared to 40 days in a conventional activated sludge reactor, probably compensated a shortening of HRT. In addition, biological assays conducted by Bertanza et al. (2011) showed that MBR was more efficient in estrogenicity reduction than a conventional activated sludge system. On the other hand, according to Clara et al. (2005b), there were no differences in BPA removal in a conventional activated sludge system and the MBR, at BPA concentrations in the influent from 720 to 2376 ng/L, at comparable SRT. Independently of the system, SRT suitable for nitrogen removal (>10 d at 10 °C) favored BPA elimination. Similar BPA removal efficiencies were found in a conventional activated sludge plant connected with a tertiary treatment consisting of flocculation-coagulation and sand filtration and two MBRs with flat sheet and hollow fibre modules, which were 80 ± 28%, 98 ± 5% and 89 ± 12%, respectively (Cases et al. 2011). In the MBR, 10–20% of initial concentration of BPA was removed during membrane filtration without contact with sludge (Seyhi et al. 2012). This could be attributed to the fact that a fraction of BPA was adsorbed on membrane surface or could be attributed to the losses of BPA owing to aeration of water.

5.1.5 BPA Removal in Full-Scale Wastewater Treatment Systems

It is necessary to optimize the design and operation of WWTPs to reduce the potential risk of micropollutant discharge into the aquatic environment. The fact that BPA was found in the effluents from WWTPs indicated that it was not completely removed during the treatment process. It is known that the efficiency of BPA removal in WWTPs depends on the type of the system and operational conditions; however, the data on the selection of the optimal technological system for BPA removal is still inconclusive. The examples of the efficiencies of BPA removal in WWTPs with different technological systems is given in Table 5.1. The results of the investigations of the fate of BPA in the subsequent stages of the treatment system are noteworthy and may have an impact on the system design. Regarding a

Table 5.1 Efficiency of BPA removal in WWTPs

Characteristics of WWTP	Efficiency of BPA removal (%)	References
Conventional activated sludge system (F/M 1.70 g COD/(g TSS d), SRT 2 d, HRT 0.08 d)	No removal	(Clara et al. 2005b)
Conventional activated sludge system (F/M 0.04 g COD/(g TSS d), SRT 46 d, HRT 1.2 d)	>95%	(Clara et al. 2005b)
Activated sludge (SRT 8–25 d, HRT 8–35 h)	87% (72 ± 10% BPA was biodegraded completely or to byproducts, 15 ± 4% was adsorbed on activated sludge, 13 ± 7% remained in the effluent)	(Stasinakis et al. 2008b)
Mechanical treatment followed by two biological treatment steps with activated sludge for degradation of organic compounds and nitrification; HRT 20–24 h	73–93%	(Höhne and Puttmann 2008)
Integrated activated sludge (anaerobic/aerobic); SRT 8–9 days; municipal wastewater (40% of the total COD load comes from the wastewater from the paper company)	89–95%	(Fürhacker et al. 2000)
Conventional activated sludge	85%	(Jacobsen et al. 2004)
Different WWTPs—conventional activated sludge, sequential batch reactor, trickling filters, aerated lagoons	1–99%	(Lee and Peart 2000)
A2O system (HRT 4–12 h, SRT 5–15 d) with tertiary treatment (flocculation and filtration)	>99%	(Sun et al. 2008)
A2O system (SRT 10–15 d, HRT 14 h)	99% in summer, 44% in winter	(Nie et al. 2012)
Activated sludge system for domestic wastewater treatment (pre-denitrification, nitrification, and secondary settling)	62%	(Bertanza et al. 2011)
Activated sludge system for domestic wastewater treatment (pre-denitrification, nitrification, and ultrafiltration)	94.5%	(Bertanza et al. 2011)

primary settling, its contribution in the removal of BPA was found negligible as BPA content in primary sludge was quite low (<10 mg/kg TSS) (Bertanza et al. 2011). A BPA-adapted bacterial community from activated sludge was able to efficiently convert BPA to CO_2 and biomass under mechanical aeration (Eio et al. 2014). A municipal WWTP employing an anaerobic/anoxic/oxic (A2O) process in which SRT was 10–15 d, HRT of the bioreactor was 14 h, and of a secondary clarifier 6 h was investigated (Nie et al. 2012). In wastewater that passed through a mechanical screen, BPA concentration was 836.9 ng/L. It increased to 994.5 ng/L when wastewater passed through an aerated grit chamber. The probable reason was that this organic compound originally adhering to the grits was peeled off due to the agitation of air. In the effluent from the secondary clarifier, the BPA concentration dropped to 3.7 ng/L. In the return sludge, BPA concentration in liquid phase was 7.6 ng/L, and in the solid phase 18.9 ng/L. In the stored excess sludge, BPA concentration in the solid phase was similar (20.1 ng/L) indicating no further biodegradation in the storage tank.

The activated sludge treatment is much more efficient than trickling filter beds for the removal of BPA, although it still does not cause its complete removal from wastewater (Kasprzyk-Hordern et al. 2009). Similarly, BPA efficiency in WWTPs with activated sludge was noted from 82 to 100%, and with trickling filters 28% (Fernandez et al. 2007). Jiang et al. (2005) compared 2 WWTPs. In WWTP 1, there was a primary clarifier, 2-stage trickling filter and 2-stage secondary clarifier. In WWTP 2, there was a primary clarifier, activated sludge tank, secondary clarifier and sand filter. BPA concentration in the influent to WWTP 2 was 2-times higher than to WWTP 1, whereas in the effluent it was significantly lower. The authors observed different efficiency of primary clarifiers: 9.4% in the WWTP 1 and 18.9% in the WWTP 2. Total efficiency of BPA removal in the WWTP with biofilm was 61.9% and with activated sludge 96%. The highest efficiency of BPA removal by activated sludge was also documented when examining rural wastewater treatment facilities which adopted activated sludge (HRT 10–16 h), constructed wetland (HRT 24–120 h), stabilization pond (HRT 24–240 h) or biofilm reactor (HRT 12–24 h) (Qiang et al. 2013). The lowest efficiency of BPA removal was obtained in the stabilization pond. Regarding technological systems with biofilm, a major trouble in moving bed biofilm reactor applications is the drop of sludge settleability when treating high strength wastewater. To solve this problem, various modifications have been developed including coagulant addition or membrane filtration as the solid separation process (Leiknes et al. 2006). For micropollutant removal, Luo et al. (2015) used a hybrid moving bed biofilm reactor–membrane bioreactor system, which could effectively remove micropollutants. As a comparison, a conventional MBR showed removal of BPA lower by 31.9%. Fouling was lower in the hybrid system than in the conventional one. In the hybrid system, the concentration of extracellular polymeric substances (EPS) was 16.24 mg/L and the concentration of soluble microbial products (SMP) was from 4.02 to 6.32 mg/L. In the conventional system, EPS concentration was 19.53 mg/L and SMP was from 21.78 to 33.04 mg/L. Although membrane fouling is often a result of organics accumulation on or in the membrane as bound EPS or SMP (Guo et al. 2012), the studies by Luo et al. (2015) indicated SMP as a main reason for membrane fouling.

BPA removal from municipal wastewater (median influent concentration 400 ng/L) was examined in WWTPs with aerated lagoons, facultative lagoons, chemically-assisted primary treatment, secondary treatment (including conventional activated sludge, trickling filters, biological aerated filters and membrane bioreactor), and advanced treatment (Guerra et al. 2015). The highest removals were observed in advanced treatment plants and the lowest in chemically-assisted primary treatment plants, in which there is no biodegradation stage and the main mechanism for pollutant removal is coagulation/flocculation and sedimentation. The authors found optimal operational conditions for BPA removal in full-scale facilities in summer and winter conditions. To gain above 80% removal during summer (median temperature 19 °C), the required conditions for HRT, SRT, and MLSS were 13 h, 7 d, and 1600 mg/L, whereas during winter (median temperature 10 °C) 13 h, 17 days, and 5300 mg/L. Therefore, to achieve good removal performance at similar HRTs, operations during cold weather require approximately 3–4 times higher SRT and MLSS concentrations than during warm weather.

5.2 Fate of BPA in Sewage Sludge

Because of the ability of BPA to sorb onto suspended solids, the mechanical and biological treatment of wastewater results in significant BPA transition from the aqueous phase to primary and secondary sludge. This means that the treated wastewater is discharged relatively free of BPA that is adsorbed on sewage sludge and which could constitute a new source of pollution. This could also cause the exposure of soil organisms to BPA if sewage sludge is applied as fertilizer on agricultural fields (Auirol et al. 2006). In addition, water solubility of BPA can increase its transport in the groundwater. However, the environmental risk assessment of BPA carried out by the European Commission showed no negative effects when an acceptable sludge amount was introduced into the soil. According to Melcer and Klečka (2011), rapid degradation of BPA occurs in soils, and therefore this compound was not detected in samples of soil fertilized with sewage sludge. A half-life of BPA in soil of less than 3 days indicated that this compound is not expected to be stable and bioavailable after disposal to soil (Fent et al. 2003).

There is a large range of concentrations of BPA in various types of sewage sludge. In the study of WWTPs with a SRT from 8 to 25 d, in which BPA removal exceeded 85%, BPA concentrations in sewage sludge did not exceed 1.75 μg/g dw (Stasinakis et al. 2008b). In raw sewage sludge from 70 to 39,800 μg BPA/kg dw was observed, in fermented sewage sludge from 33 to 36,700 μg BPA/kg dw. The highest BPA concentration (325,000 μg/kg dw) was noted in the sludge coming from a municipal WWTP with 50% contribution of industrial wastewater (Meesters and Schroder 2002). BPA was also detected in excess sludge at concentration of 0.26 mg/kg TSS; however, mass balance showed that the amount found in excess sludge accounted for less than 0.5% of the mass entering the biological system (Bertanza et al. 2011). BPA concentrations were higher in primary sludge than in secondary sludge (Weltin et al. 2002).

There is little data on the fate of BPA during sewage sludge treatment. The impact of the methods of sludge conditioning on BPA concentration in the sludge was analyzed (Ivashechkin et al. 2004). The application of organic polymers or inorganic coagulants (iron chloride (III)) did not affect the effectiveness of the BPA sorption in sewage sludge while the use of calcium hydroxide resulted in enhanced desorption of BPA. A ca. two times increase in the concentration of BPA in anaerobically digested biosolids was observed as compared to primary and secondary sludge (Guerra et al. 2015). The potential release of BPA during methane fermentation could lead to an increase in the load of BPA in the aeration tanks as a result of the recirculation of reject water. Observed accumulation of BPA in anaerobically digested biosolids resulted from a substantial reduction of volatile solids during digestion and from slight BPA degradation under anaerobic conditions (Mohapatra et al. 2010b). Compared to anaerobic digestion, lower BPA levels were observed in non-digested biosolids and in biosolids from aerobic digestion, probably due to degradation processes (Mohapatra et al. 2010b). BPA concentration in dewatered sludge after mesophilic anaerobic sludge digestion from municipal WWTP was 1.86 mg/kg and dropped to 0.11 mg/kg after thermal drying (Samaras et al. 2013); the decrease was not explained.

The fate of BPA in sewage sludge can be affected by its rheological characteristics. When physico-chemical processes like e.g. ultrasonication, Fenton oxidation or hydrolysis are used as a pre-treatment step, they change the surface chemistry of sewage sludge, particularly proteins and polysaccharides, as well as their particle size (Stasinakis 2008; Pham et al. 2009). These changes affect the viscosity, which influences mass transfer, thus affecting biodegradation and sorption/desorption processes (Mohapatra et al. 2010a).

Apart from the commonly used handling, processing and disposal of sewage sludge, its re-use to produce value-added products (biofertilizer, biopesticides, bioherbicides, bioplastics and biocontrol agents) is a novel approach of sewage sludge management (Mohapatra et al. 2010a). During bioconversion of sewage sludge into value-added products, commercial microorganisms secrete enzymes that have improved access to the organic compounds so they can simultaneously detoxify or degrade toxic compounds. Therefore, it is crucial to select microorganism cultivable in sludge to concomitant production value-added products and enzymes to detoxify or degrade BPA. As value-added producers, *Trichoderma* sp., *Rhizobium* sp. and *Bacillus* sp. showed high potential to grow in wastewater and sewage sludge, produce microbial derivatives used for biocontrol and bio-bleaching processes and possess the enzyme system which can efficiently degrade BPA (Verma et al. 2007; Brar et al. 2008). Particularly, *Trichoderma viride* is a promising microorganism which produces laccases and could potentially degrade or detoxify BPA (Verma et al. 2007). Due to the fact that the accumulation of BPA in sewage sludge questions the benefits of production of these products, BPA removal from sewage sludge is necessary, including determination of the fate not only BPA but also its toxic intermediates that may be generated. Therefore, further studies to determine how the generation of value-added products affects the fate of BPA and its by-products in the sludge are necessary.

The above-mentioned data indicate that wastewater treatment plant is a source of BPA in the environment both by discharged wastewater and sewage sludge. In addition, the application of sewage sludge from WWTPs that receive BPA-containing wastewaters is a potential route of entry of BPA to the soil.

References

Abargues MR, Ferrer J, Bouzas A et al (2013) Removal and fate of endocrine disruptors chemicals under lab-scale postreatment stage. Removal assessment using light, oxygen and microalgae. Bioresour Technol 149:142–148

Aguayo S, Munoz MJ, de la Torre A et al (2004) Identification of organic compounds and ecotoxicological assessment of sewage treatment plants (STP) effluents. Sci Total Environ 328:69–81

Alpaslan Kocamemi B, Cecen F (2007) Inhibitory effect of the xenobiotic 1,2-DCA in a nitrifying biofilm reactor. Water Sci Technol 55:67–73

Alvarez-Cohen L, McCarty PL (1991) Effects of toxicity, aeration, and reductant supply on TCE transformation by a mixed methanotrophic culture. Appl Environ Microbiol 57:228–235

Auirol M, Filali-Meknassi Y, Tyagi RD et al (2006) Endocrine disrupting compounds removal from wastewater: a new challenge. Process Biochem 41:525–539

Balest L, Lopez A, Mascolo G et al (2008) Removal of endocrine disrupter compounds from municipal wastewater using an aerobic granular biomass reactor. Biochem Eng J 41:288–294

Bertanza G, Pedrazzani R, Dal Grande M et al (2011) Effect of biological and chemical oxidation on the removal of estrogenic compounds (NP and BPA) from wastewater: an integrated assessment procedure. Water Res 45:2473–2484

Boonnorat J, Chiemchaisri C, Chiemchaisri W et al (2014) Removals of phenolic compounds and phthalic acid esters in landfill leachate by microbial sludge of two-stage membrane bioreactor. J Hazard Mater 277:93–101

Boonyaroj V, Chiemchaisri C, Chiemchaisri W et al (2012) Toxic organic micro-pollutants removal mechanisms in long-term operated membrane bioreactor treating municipal solid waste leachate. Bioresour Technol 113:174–180

Brar SK, Verma M, Tyagi RD et al (2008) *Bacillus thuringiensis* fermentation of wastewater and wastewater sludge—presence and characterization of chitinases. Environ Technol 29:161–170

Cabana H, Jiwan JLH, Rozenberg R et al (2007) Elimination of endocrine disrupting chemicals nonylphenol and bisphenol A and personal care product ingredient triclosan using enzyme preparation from the white rot fungus *Coriolopsis* polyzona. Chemosphere 67:770–778

Cases V, Alonso V, Argandoña V et al (2011) Endocrine disrupting compounds: a comparison of removal between conventional activated sludge and membrane bioreactors. Desalination 272:240–245

Chen J, Huang X, Lee D (2008) Bisphenol A removal by a membrane bioreactor. Process Biochem 43:451–456

Chong NM, Chen YS (2007) Activated sludge treatment of a xenobiotic with or without a biogenic substrate during start-up and shocks. Bioresour Technol 98:3611–3616

Chong NM, Lin TY (2007) Measurement of the degradation capacity of activated sludge for a xenobiotic organic. Bioresour Technol 98:1124–1127

Clara M, Strenn B, Saracevic E et al (2004) Adsorption of bisphenol A, 17β-ethinylestradiole to sewage sludge. Chemosphere 56:843–851

Clara M, Kreuzinger N, Strenn B et al (2005a) The solids retention time—a suitable design parameter to evaluate the capacity of wastewater treatment plants to remove micropollutants. Water Res 39:97–106

Clara M, Strenn B, Gans O et al (2005b) Removal of selected pharmaceuticals, fragrances and endocrine disrupting compounds in a membrane bioreactor and conventional wastewater treatment plants. Water Res 39:4797–4807

Daims H, Purkhold U, Bjerrum L et al (2001) Nitrification in sequencing biofilm batch reactors: lessons from molecular approaches. Water Sci Technol 43:9–18
Demarche P, Junghanns C, Mazy N et al (2012a) Design-of-experiment strategy for the formulation of laccase biocatalysts and their application to degrade bisphenol A. New Biotechnol 30:96–103
Demarche P, Junghanns C, Nair RR et al (2012b) Harnessing the power of enzymes for environmental stewardship. Biotech Adv 30:933–953
De Wever H, Weiss S, Reemtsma T et al (2007) Comparison of sulfonated and other micropollutants removal in membrane bioreactor and conventional wastewater treatment. Water Res 41:935–945
Drewes JE, Hemming J, Ladenburger DJ et al (2005) An assessment of endocrine disrupting activity changes during wastewater treatment through the use of bioassays and chemical measurements. Water Environ Res 77:12–23
Eio EJ, Kawai M, Tsuchiya K et al (2014) Biodegradation of bisphenol A by bacterial consortia. Int Biodeterior Biodegrad 96:166–173
Falas P, Baillon-Dhumez A, Andersen HR et al (2012) Suspended biofilm carrier and activated sludge removal of acidic pharmaceuticals. Water Res 46:1167–1175
Fatone F, Bolzonella D, Battistoni P et al (2005) Removal of nutrients and micropollutants treating low loaded wastewaters in a membrane bioreactor operating the automatic alternate-cycles process. Desalination 183:395–405
Felis E, Borok S, Miksch K (2011) The evaluation of the selected hormonal biomimetics for sorption on the flocs of activated sludge (Ocena zdolności wybranych biomimetyków hormonalnych do sorpcji na kłaczkach osadu czynnego). Ochrona Środowiska 33:49–52 (In Polish language)
Fent G, Hein WJ, Moendel M et al (2003) Fate of ^{14}C-bisphenol A in soils. Chemosphere 51:735–746
Fernandez MP, Ikonomou MG, Buchanan I (2007) Assessment of estrogenic organic contaminants in Canadian wastewater. Sci Total Environ 373:250–269
Fernandez MP, Noguerol T, Lacorte S et al (2009) Toxicity identification fractionation of environmental estrogens in waste water and sludge using gas and liquid chromatography coupled to mass spectrometry and recombinant yeast assay. Anal Bioanal Chem 393:957–968
Fernandez-Fontaina E, Omil F, Lema JM et al (2012) Influence of nitrifying conditions on the biodegradation and sorption of emerging micropollutant. Water Res 46:5434–5444
Ferro Orozco AM, Lobo CC, Contreras EM et al (2013) Biodegradation of bisphenol-A (BPA) in activated sludge batch reactors: analysis of the acclimation process. Int Biodeter Biodegr 85:392–399
Fürhacker M, Scharf S, Weber H (2000) Bisphenol A: emissions from point sources. Chemosphere 41:751–756
Guerra P, Kim M, Teslic S et al (2015) Bisphenol-A removal in various wastewater treatment processes: operational conditions, mass balance, and optimization. J Environ Manag 152:192–200
Guo W, Ngo HH, Li J (2012) A mini-review on membrane fouling. Bioresour Technol 122:27–34
Höhne C, Puttmann W (2008) Occurrence and temporal variations of the xenoestrogens bisphenol A, 4-tert-octylphenol, and tech. 4-nonylphenol in two German wastewater treatment plants. Environ Sci Pollut Res Int 15:405–416
Ike M, Chen MY, Danzl E et al (2006) Biodegradation of a variety of bisphenols under aerobic and anaerobic conditions. Water Sci Technol 53:153–159
Ivashechkin P, Corvini PFX, Dohmann M (2004) Behaviour of endocrine disrupting chemicals during the treatment of municipal sewage sludge. Water Sci Technol 50:133–140
Jacobsen BN, Kjersgaard D, Winther-Nielsen M et al (2004) Combined chemical analyses and biomonitoring at avedoere wastewater treatment plant in 2002. Water Sci Technol 50:37–43
Jewell KS, Wick A, Ternes TA (2014) Comparisons between abiotic nitration and biotransformation reactions of phenolic micropollutants in activated sludge. Water Res 48:478–489

Jiang JQ, Yin Q, Zhou JL et al (2005) Occurrence and treatment trials of endocrine disrupting chemicals (EDCs) in wastewaters. Chemosphere 61:544–550

Joss A, Zabczynski S, Gobel A et al (2006) Biological degradation of pharmaceuticals in municipal wastewater treatment: proposing a classification scheme. Water Res 40:1686–1696

Kang JH, Kondo F (2002a) Effects of bacterial counts and temperature on the biodegradation of bisphenol A in river water. Chemosphere 49:493–498

Kang JH, Kondo F (2002b) Bisphenol A degradation by bacteria isolated from river water. Arch Environ Contam Toxicol 43:265–269

Kasprzyk-Hordern B, Dinsdale RM, Guwy AJ (2009) The removal of pharmaceuticals, personal care products, endocrine disruptors and illicit drugs during wastewater treatment and its impact on the quality of receiving waters. Water Res 43:363–380

Kim YI, Nicell JA (2006) Impact of reaction conditions on the laccase-catalysed conversion of bisphenol A. Bioresour Technol 97:1431–1442

Kim S, Eichhorn P, Jensen JN et al (2005) Removal of antibiotics in wastewater: effect of hydraulic and solid retention times on the fate of tetracycline in the activated sludge process. Environ Sci Technol 39:5816–5823

Kim JY, Ryu K, Kim EJ et al (2007) Degradation of bisphenol A and nonylphenol by nitrifying activated sludge. Process Biochem 42:1470–1474

Klečka GM, Gonsior SJ, West RJ et al (2001) Biodegradation of bisphenol A in aquatic environments: river die-away. Environ Toxicol Chem 20:2725–2735

Koponen PS, Kukkonen JVK (2001) Effects of bisphenol A and artificial UVB radiation on the early development of Rana temporaria. J Toxicol Environ Health A 65:947–959

Kovarova-Kovar K, Egli T (1998) Growth kinetics of suspended microbial cells: from single-substrate controlled growth to mixed-substrate kinetics. Microbiol Mol Biol Rev 62:646–666

Langford K, Scrimshaw M, Lester J (2007) The impact of process variables on the removal of PBDEs and NPEOs during simulated activated sludge treatment. Arch Environ Contam Toxicol 53:1–7

LaPara TM, Nakatsu CH, Pantea LM et al (2002) Stability of the bacterial communities supported by a seven-stage biological process treating pharmaceutical wastewater as revealed by PCR-DGGE. Water Res 36:638–646

LaPara TM, Klatt CG, Chen R (2006) Adaptations in bacterial catabolic enzyme activity and community structure in membrane-coupled bioreactors fed simple synthetic wastewater. J Biotechnol 121:368–380

Lee HB, Peart TE (2000) Bisphenol A contamination in Canadian municipal and industrial wastewater and sludge samples. Water Qual Res J Can 35:283–298

Leiknes T, Bolt H, Engmann M et al (2006) Assessment of membrane reactor design in the performance of a hybrid biofilm membrane bioreactor (BFMBR). Desalination 199:328–330

Li J, Jiang L, Liu X et al (2013) Adsorption and aerobic biodegradation of four selected endocrine disrupting chemicals in soil-water system. Int Biodeter Biodegr 76:3–7

Luo Y, Jiang Q, Ngo HH et al (2015) Evaluation of micropollutant removal and fouling reduction in a hybrid moving bed biofilm reactor–membrane bioreactor system. Bioresour Technol 191:355–359

Meesters RJW, Schroder HF (2002) Simultaneous determination of 4-nonylphenol and bisphenol A in sewage sludge. Anal Chem 74:3566–3574

Melcer H, Klečka G (2011) Treatment of wastewaters containing bisphenol A: state of the science review. Water Environ Res 83:650–666

Miserez K, Philips S, Verstraete W (1999) New biology for advanced wastewater treatment. Water Sci Technol 40:137–144

Mohapatra DP, Brar SK, Tyagi RD et al (2010a) Degradation of endocrine disrupting bisphenol A during pre-treatment and biotransformation of wastewater sludge. Chem Eng J 163:273–283

Mohapatra DP, Brar SK, Tyagi RD et al (2010b) Physico-chemical pretreatment and biotransformation of wastewater and wastewater sludge—fate of bisphenol A. Chemosphere 78:923–941

Nakajima N, Teramoto T, Kasai F et al (2007) Glycosylation of bisphenol A by freshwater microalgae. Chemosphere 69:934–941
Nasu M, Goto M, Kato H et al (2001) Study on endocrine disrupting chemicals in wastewater treatment plants. Water Sci Technol 43:101–108
Nghiem LD, Tadkaew N, Sivakumar M (2009) Removal of trace organic contaminants by submerged membrane bioreactors. Desalination 236:127–134
Nie Y, Qiang Z, Zhang H et al (2012) Fate and seasonal variation of endocrine-disrupting chemicals in a sewage treatment plant with A/A/O process. Sep Purif Technol 84:9–15
Ogawa H, Kitamura H, Miyata N (2005) Biodegradation of endocrine disrupting chemicals in aerobic and anaerobic sludges. Jpn J Water Treat Biol 41:83–92
Pathak A, Dastidar MG, Sreekrishnan TR (2009) Bioleaching of heavy metals from sewage sludge: a review. J Environ Manag 90:2343–2353
Pham TTH, Brar SK, Tyagi RD et al (2009) Ultrasonication of WWS-consequences on biodegradability and flowbility. J Hazard Mater 163:2–13
Pothitou P, Voutsa D (2008) Endocrine disrupting compounds in municipal and industrial wastewater treatment plants in Northern Greece. Chemosphere 73:1716–1723
Press-Kristensen K, Lindblom E, Schmidt JE et al (2008) Examining the biodegradation of endocrine disrupting bisphenol A and nonylphenol in WWTPs. Water Sci Technol 57:1253–1256
Qiang Z, Dong H, Zhu B et al (2013) A comparison of various rural wastewater treatment processes for the removal of endocrine-disrupting chemicals (EDCs). Chemosphere 92:986–992
Ren YX, Nakano K, Nomura M et al (2007) Effects of bacterial activity on estrogen removal in nitrifying activated sludge. Water Res 41:3089–3096
Roh H, Subramanya N, Zhao F et al (2009) Biodegradation potential of wastewater micropollutants by ammonia-oxidizing bacteria. Chemosphere 77:1084–1089
Samaras VG, Stasinakis AS, Mamais D et al (2013) Fate of selected pharmaceuticals and synthetic endocrine disrupting compounds during wastewater treatment and sludge anaerobic digestion. J Hazard Mater 244–245:259–267
Sarmah AK, Northcott GL (2008) Laboratory degradation studies of four endocrine disruptors in two environmental media. Environ Toxicol Chem 27:819–827
Sasaki M, Maki JI, Oshiman KI et al (2005) Biodegradation of bisphenol A by cells and cell lysate from *Sphingomonas* sp. strain AO1. Biodegradation 16:449–459
Schwarzenbach RP, Gschwend PM, Imboden DM (2003) Environmental organic chemistry. Wiley, Hoboken
Seyhi B, Drogui P, Buelna G et al (2011) Modeling of sorption of bisphenol A in sludge obtained from a membrane bioreactor process. Chem Eng J 172:61–67
Seyhi B, Drogui P, Buelna G et al (2012) Removal of bisphenol-A from spiked synthetic effluents using an immersed membrane activated sludge process. Sep Purif Technol 87:101–109
Shen G, Yu G, Cai ZX et al (2005) Development of an analytical method to determine phenolic endocrine disrupting chemicals in sewage and sludge by GUMS. Chin Sci Bull 50:2681–2687
Speitel GE, Segar RL (1995) Cometabolism in biofilm reactors. Water Sci Technol 31:215–225
Spivack J, Leib TK, Lobos JH (1994) Novel pathway for bacterial metabolism of bisphenol A. J Biol Chem 269:7323–7329
Staples CA, Dorn PB, Klečka GM et al (1998) A review of the environmental fate, effects and exposures of bisphenol A. Chemosphere 36:2149–2173
Stasinakis AS (2008) Use of selected advanced oxidation processes (AOPs) for wastewater treatment—a mini review. Global NEST J 10:376–385
Stasinakis AS, Petalas AV, Mamais D et al (2008a) Application of the OECD 301F respirometric test for the biodegradability assessment of various potential endocrine disrupting chemicals. Bioresour Technol 99:3458–3467
Stasinakis AS, Gatidou G, Mamais D et al (2008b) Occurrence and fate of endocrine disrupters in Greek sewage treatment plants. Water Res 42:1796–1804

Stasinakis AS, Kordoutis CI, Tsiouma VC et al (2010) Removal of selected endocrine disrupters in activated sludge systems: effect of sludge retention time on their sorption and biodegradation. Bioresour Technol 101:2090–2095

Stringfellow WT, Alvarez-Cohen L (1999) Evaluating the relationship between the sorption of PAHs to bacterial biomass and biodegradation. Water Res 33:2535–2544

Sun Q, Deng S, Huang J et al (2008) Contributors to estrogenic activity in wastewater from a large wastewater treatment plant in Beijing, China. Environ Toxicol Phar 25:20–26

Suzuki T, Nakagawa Y, Takano I et al (2004) Environmental fate of bisphenol A and its biological metabolites in river water and their xeno-estrogenic activity. Environ Sci Technol 38:2389–2396

Tadkaew N, Hai FI, McDonald JA et al (2011) Removal of trace organics by MBR treatment: the role of molecular properties. Water Res 45:2439–2451

Toyoizumi T, Deguchi Y, Masuda S et al (2008) Genotoxicity and estrogenic activity of 3,30-dinitrobisphenol a in goldfish. Biosci Biotechnol Biochem 72:2118–2123

Urase T, Kikuta T (2005) Separate estimation of adsorption and degradation of pharmaceutical substances and estrogens in the activated sludge process. Water Res 39:1289–1300

Verma M, Brar SK, Tyagi RD et al (2007) Bench-scale fermentation of *Trichoderma viride* on WWS: rheology, lytic enzymes and biocontrol agents. Enzyme Microb Technol 41:764–771

Verstraete W, Philips S (1998) Nitrification-denitrification processes and technologies in new context. Environ Pollut 102:717–726

Weltin D, Gehring M, Tennhardt L et al (2002) Occurrence and fate of bisphenol A during wastewater treatment in selected German sewage treatment plants. In: Proceedings of the 2002 American water works association conference, endocrine disruptors and the water industry, American water works association, Denver, Colorado, 18–20 April 2002

West RJ, Goodwin PA, Klečka GM (2001) Assessment of the ready biodegradability of Bisphenol A. Bull Environ Toxicol Chem 67:106–112

Wintgens T, Gallenkemper M, Melin T (2004) Removal of endocrine disrupting compounds with membrane processes in wastewater treatment and reuse. Water Sci Technol 50:1–8

Wood PM (1990) Autotrophic and heterotrophic mechanisms for ammonia oxidation. Soil Use Manag 6:78–79

Xia S, Li J, Wang R (2008) Nitrogen removal performance and microbial community structure dynamics response to carbon-nitrogen ratio in a compact suspended carrier biofilm reactor. Ecol Eng 32:256–262

Xiangli Q, Zhenjia Z, Quingxuan C et al (2008) Nitrification characteristics of PEG immobilized activated sludge at high ammonia and COD loading rates. Desalination 222:340–347

Xie Y, Li H, Wang L et al (2011) Molecularly imprinted polymer microspheres enhanced biodegradation of Bisphenol A by acclimated activated sludge. Water Res 45:1189–1198

Yuan SY, Shiung LC, Chang BV (2002) Biodegradation of polycyclic aromatic hydrocarbons by inoculated microorganisms in soil. Bull Environ Contam Toxicol 69:66–73

Zhao J, Li Y, Zhang C et al (2008) Sorption and degradation of bisphenol A by aerobic activated sludge. J Hazard Mater 155:305–311

Zhao G, Zhou L, Li Y et al (2009) Enhancement of phenol degradation using immobilized microorganisms and organic modified montmorillonite in a two-phase partitioning bioreactor. J Hazard Mater 169:402–410

Zhou NA, Lutovsky AC, Andaker GL et al (2014) Kinetics modeling predicts bioaugmentation with Sphingomonad cultures as a viable technology for enhanced pharmaceutical and personal care products removal during wastewater treatment. Bioresour Technol 166:58–167

Zielińska M, Cydzik-Kwiatkowska A, Bernat K et al (2014) Removal of bisphenol A (BPA) in a nitrifying system with immobilized biomass. Bioresour Technol 171:305–313

Chapter 6
Integrated Systems for Removal of BPA from Wastewater

There is an increasing interest in the combined methods of BPA removal from wastewater that involve mostly the integration of chemical and biological approaches, in which the conventional biological treatment is supported by Advanced Oxidation Processes (AOPs) such as photocatalysis, ozone oxidation, Fenton and photo-Fenton oxidation, and wet air oxidation (WAO). These chemical treatments, used to break down BPA to molecules suitable for biotreatment, have been developed mostly to improve the recycling and re-use of wasted sludge. The pre-treatment enhances the hydrolysis of sludge, which reduces the stabilization time and increases the degree of degradation during biological treatment. Wasted sludge pre-treatments rupture suspended solids (microbial cells), liberate the nutrients, partially solubilize the suspended solids, increase the soluble chemical oxygen demand and decrease sludge viscosity. In addition, the utilization of membranes for polishing biological secondary effluent and for supporting AOPs has been reported. These integrated methods have been applied to increase the biodegradability, improve the quality and detoxify the effluent streams. They are considered advantageous for water reuse that requires high efficiency of pollutant removal. This chapter aims at providing a review on the integration of different treatment approaches which have been described as favorable for the acquisition of high-quality effluent.

6.1 Technologies Combining AOPs and Biological Treatment

Among integrated treatment processes, the combination of conventional biotreatment with AOPs is reported as a highly effective system used for purifying recalcitrant wastewater and recycling and re-using wasted sludge contaminated by endocrine disrupting compounds (EDCs), including BPA. The AOPs are used as a

M. ZIELIŃSKA et al., *Bisphenol A Removal from Water and Wastewater*,
https://doi.org/10.1007/978-3-319-92361-1_6

pre-treatment step to convert the persistent BPA into more biodegradable intermediates, which would then be treated in a biological oxidation process. The percentage of mineralization should be minimal during the pre-treatment to avoid unnecessary expenditure of chemicals and energy, thereby lowering the operating cost. This is important because electricity represents about 60% of the total operating cost of the photocatalytic reactors. If the pre-treatment time is too short, the generated intermediates could still be structurally very similar to the original non-biodegradable and/or toxic components. It is very important to perform toxicity analyses (with organisms like *Vibrio fischeri*, *Daphnia magna*, activated sludge by respirometric assays, etc.) and biodegradability tests (using activated sludge) to assess if the AOP effluent is appropriate to be subsequently treated by a conventional biodegradation process.

Photocatalysis is a chemical oxidation process in which a metal oxide semiconductor is immersed in water and irradiated by near UV light ($\lambda < 385$ nm) that results in the formation of free hydroxyl ($^{\cdot}OH$) radicals (Belgiorno et al. 2007). Although several semiconductors exist, TiO_2 is the most widely used, because of its photo-stability, non-toxicity, low cost and water insolubility. Photocatalysis has been reported to be effective for the degradation of BPA in wastewater. 63% removal of BPA was obtained at its initial concentration of 100 mg/L, at pH 6, 1 h reaction time and TiO_2 concentration of 10 g/L (Kaneco et al. 2004). 95–99% of BPA removal was noted at its initial concentration of 20 mg/L, 1 h reaction time and TiO_2 concentration of 2 g/L (Chiang et al. 2004). These authors showed a strong dependence of BPA mineralization on the pH. BPA was completely mineralized at pH 3.0 after 120 min of UV irradiation, whereas oxidation intermediates formed at pH 10 were less toxic compared to the BPA molecule. Apart from TiO_2 as a catalyst, Coleman et al. (2005) investigated the effects of silver and platinum metals on the photocatalytic degradation of BPA. There was no effect of the addition of silver or platinum on the mineralization efficiency; hence they concluded that the use of expensive metals in photocatalytic systems for the removal of low EDCs concentrations is unjustifiable.

Photodegradation of BPA is enhanced by algae. The algae may produce some secretions which after irradiation can produce hydroxyl radicals that can enhance the BPA degradation (Peng et al. 2006).

To overcome the problem of large amounts of excess sludge produced in WWTPs, novel technologies of excess sludge reduction have been developed in recent years, among which ozonation is considered to be an effective and practical method. The sludge reduction efficiency by ozonation was reported as reaching 100%, which means no production of excess sludge (Lee et al. 2005). However, sludge ozonation may cause some hydrophobic organic micro-pollutants adsorbed on sludge to get released into wastewater because of sludge solubilization and thus contaminate the receiving water bodies (Liu 2003). The effect of sludge ozonation on the degradation of several EDCs including BPA was examined (Qiang et al. 2013). BPA present in activated sludge could be effectively removed by O_3 although the rate of degradation was 3–4 orders of magnitude lower than those in water. The authors indicate the necessity of adjusting an appropriate dose of O_3

because a low dose may results in an increase of BPA concentration in the liquid phase of activated sludge. In addition, tertiary ozonation had the beneficial effect on the reduction of estrogenicity of the effluent of conventional activated sludge system for domestic wastewater treatment and BPA removal (Bertanza et al. 2011). Rate constants were dependent on reagent dosage; a 90% removal of BPA was achieved either after 80 min at 8 mg O_3/L or after 27 min at 11 mg O_3/L. The combination of ozonation and sand filtration with activated sludge treatment gave EDC removal >80% (Nakada et al. 2007).

The effects of different pre-treatment methods, including ultrasonication, Fenton's oxidation and ferro-sonication were assessed in terms of increased solubilization of wastewater sludge and simultaneous degradation of BPA. Among these three pre-treatment methods, the highest removal of total suspended solids, volatile suspended solids, chemical oxygen demand and soluble organic carbon was obtained during a ferro-sonication carried out for 180 min, resulting in high efficiency of BPA degradation. Efficient removal of BPA in the process was a result of the high activity of laccases produced by *Sinorhizobium meliloti* (Mohapatra et al. 2011a). The application of ultrasound is considered an innovative method of treatment of hazardous chemicals such as BPA (Torres et al. 2006; Belgiorno et al. 2007). Destruction of micropollutants is usually achieved through a combination of pyrolytic reactions occurring inside or near the cavitation bubble and hydroxyl radical-mediated reactions occurring in the liquid bulk (Papadaki et al. 2004). It was found that ultrasonication time significantly affected biodegradation of BPA; maximum efficiency of 82.7% was obtained with 180 min ultrasonication time (Mohapatra et al. 2011b).

For sewage sludge pre-treatment prior to methane fermentation, Mohapatra et al. (2010) used alkaline hydrolysis, thermal hydrolysis, thermal alkaline hydrolysis, thermal oxidation and thermal alkaline oxidation. The order of efficiency of pre-treatment methods for solubilisation of sewage sludge was thermal alkaline hydrolysis > thermal alkaline oxidation > thermal oxidation > thermal hydrolysis > alkaline hydrolysis, whereas that for BPA degradation was thermal alkaline oxidation > thermal oxidation > thermal alkaline hydrolysis > thermal hydrolysis > alkaline hydrolysis. As compared to alkaline pre-treated sludge, higher degradation of BPA was observed after thermal pre-treatment, which leads to lysis of sludge, hydration of macromolecules secreted from disrupted sludge cells and the transfer of organic compounds from solid to liquid phase. Among 3 processes based on hydrolysis leading to breaking the sludge solid fraction into soluble and less complex molecules, thermal alkaline hydrolysis improved sewage sludge solubilisation and BPA biodegradation the most. It has been shown that the disintegration of sewage sludge as a result of this process positively influenced the BPA biodegradation by laccase produced by bacterium *Sinorhizobium meliloti.* This has the potential to degrade or detoxify organic compounds, especially with phenolic groups. However, the highest degree of BPA degradation was detected during oxidation of sludge (75.0 and 78.9%) that is characterized by the generation of hydroxyl radicals. These radicals are one of the strongest known oxidants for oxidation and mineralization of organic molecules. During these oxidation processes, three intermediates were formed: hydroquinone, 4-hydroxyacetophenone and 3-hydroxybisphenol A. The reaction of these intermediates with hydroxyl radicals leads to formation of low-weight hydrocarbons such as formic acid, propionic acid, formaldehyde and further CO_2 and H_2O (Torres et al. 2007). The disintegration methods used by Mohapatra et al. (2010) resulted in a decrease in viscosity and particle sizes of the sludge, which supported BPA degradation.

During thermal pre-treatment, the waste activated sludge is generally subjected to temperatures in the range from 140 to 250 °C, with corresponding pressures in the range of 0.6 to 2.5 MPa. Sub-critical wet oxidation is conducted at higher temperatures (150–370 °C) and pressures (5–30 MPa) than thermal hydrolysis (Genc et al. 2002). The crucial difference between thermal hydrolysis (TH) and wet oxidation (WO) is that TH relies on heat to destroy the bonds within organic material, while WO makes use of both heat and an oxidant (usually oxygen) to degrade organic compounds. Depending on the severity of the conditions, the ultimate end-product of the wet oxidation of sludge is CO_2. The use of an oxidant allows for higher reaction rates and efficiency of destruction and produces different residual compounds. These compounds are often dominated by a range of short-chain fatty acids such as acetate, formate, propionate and valerate. During thermal pre-treatment, the optimum anaerobic digestion (AD) conditions and improvement in methane synthesis varies considerably and is dependent on the nature of the sludge combination of thermal/thermo-chemical and biological treatment, summarized by the following model (Strong and Gapes 2012):

$$\text{Waste material} + \text{TH/WO} \rightarrow \text{Solid residue} + \text{dissolved organic carbon} - \text{AD} \\ \rightarrow CH_4 + CO_2 + \text{final residue}$$

WO results in a rapid destruction of municipal waste activated sludge. The destruction of 83% TSS and 93% VSS was higher than could be achieved using thermal hydrolysis, anaerobic digestion or combinations thereof. WO may not, however, be optimal for methane generation due to the destruction of solubilised organic compounds. During WO of sludge, a large proportion of the insoluble organics are first solubilised through hydrolytic depolymerisation reactions which, together with the formation of free radicals, are the most important oxidising species. Subsequent oxidative reactions convert these hydrolysis products into increasingly simpler organics such as organic acids and alcohols. Finally, these products can be further oxidised to CO_2, N_2, water and residual ash. The mechanism of the oxidation process based on free radical chain reactions includes (i) initiation—reaction of molecular oxygen with organic compounds (RH) to produce alkyl ($R^{\cdot}$) and hydroperoxyl ($ROO^{\cdot}$) radicals, or production of alkyl ($R^{\cdot}$) and hydrogen ($H^{\cdot}$) radicals through non-oxidative hydrothermal reactions; (ii) propagation—further reactions of oxygen and other organic and inorganic radicals to produce strong oxidant hydrogen peroxide (H_2O_2); and (iii) oxidation—adding organic acid links (OOH) to the organic fragments which results in the formation of organic acids (Baroutian et al. 2016). It could be a way to treat wastewater sludge with BPA.

6.2 Technologies Combining AOPs and Membrane Filtration

One of the attempts to enhance oxidation and remove BPA was the use of membranes. Membrane separation is becoming a very attractive alternative because of purely physical nature of separation as well as the modular design of membrane processes. The using of only membrane filtration for wastewater purification is limited by the clogging of membranes with pollutants (Sun et al. 2015), which causes the shortening of filtration cycle and lowering of membrane life. So called fouling accompanies membrane filtration and is caused by the presence of organic compounds in wastewater. This may affect the removal of BPA. To improve the performance of membrane for BPA retention and to lengthen membrane life, different hybrid processes can be used to remove BPA and other EDCs. These hybrid processes combine membrane filtration with e.g. Fenton's process, adsorption or other techniques to enhance BPA retention or to achieve simultaneous retention and mineralization. However, significant membrane fouling was observed in the nanofiltration of the effluent even after Fenton oxidation (Escalona et al. 2014). Furthermore, these hybrid processes for BPA removal are only applied in laboratory scale at present stage (Liang et al. 2015).

Although TiO_2 is considered a good photocatalyst due to its photoactivity, physical and chemical stability, it is almost impossible to efficiently separate the catalyst from the liquid stream by conventional solid-liquid separation techniques. Thus, the recovery of nanoparticles of this photocatalysts from the liquid remains a major engineering problem to be solved before the catalyst can be used for large-scale applications (Kim et al. 2008b). Immobilizing particulate TiO_2 on a substrate solved this problem but, on the other hand, catalyst immobilization diminishes mass transport (Butterfield et al. 1997). Among different attempts to improve the recovery of particulate TiO_2, membrane technology was believed to improve the implementation of the photocatalytic oxidation process by enhancing the separation of particulate TiO_2 from the treated water. This membrane technology utilized high-pressure NF (Molinari et al. 2002), which retains low molecular substances, and low-pressure membranes such as MF and UF (Chin et al. 2007), which have an energy benefit over NF and can be used as submerged membranes. A hybrid system combining a low-pressure submerged hollow fibre membrane module being in direct contact with the photocatalytic oxidation medium to treat water containing BPA was used (Chin et al. 2007). Membranes were used for retaining the TiO_2 particles in the system. A combination of pH 4, 0.5 g TiO_2/L and aeration rate of 0.5 L/min was found to be the optimal conditions, in which 97% of BPA was photodegraded and 93% of TOC was photomineralized. Aeration in a photocatalytic reactor enhanced the photodegradation rate by providing higher concentrations of dissolved oxygen and improving the mass transfer process in the system. Photoactivity was found to increase with aeration rate up to a limit of 0.5 L/min. It was found that intermittent filtration reduced fouling through the removal of the TiO_2 layer from the membrane surface and thus allowed high flux operation, enhanced the sustainability of the submerged membranes but showed no effect on the photoactivity of the system. In the studies by Kim et al. (2008b), a hybrid system for BPA removal from drinking water combined NF with homogeneous catalytic oxidation, in which iron(III)–tetrasulfophthalocyanine was used as a homogeneous metal catalyst in the presence of hydrogen peroxide. BPA oxidation resulted in over 90% decomposition of BPA within 3 min under weakly acidic conditions (pH $\leq$ 4.5). By-products of BPA oxidation were *p*-benzoquinone, 4-isopropenyl phenol, BPA-*o*-quinone, and 4-hydroxyphenyl-2-propanol. These by-products could further be converted to low molecular weight organic acids. The catalyst was retained in the system by the NF and used to continuously decompose BPA. As a result, the hybrid system with NF showed higher efficiency of BPA removal (95%) than the NF-only system (72%). However, in the hybrid system with NF permeate flux was slightly lower due to precipitation of the catalyst on the membrane surface. In a similar technology, the catalyst iron(III)-tetrasulfophthalocyanine was electrostatically immobilized on anion-exchange resin to improve catalyst stability (Kim et al. 2008a). The results of this integrated system showed complete removal of BPA by adsorption and oxidation in the presence of hydrogen peroxide, even though anion-exchange resin itself could adsorb some amount of BPA. The immobilized catalyst increased BPA removal to over 98% within 1 h at a neutral pH.

BPA degradation by Fenton's process and subsequent NF of low concentration of residual BPA and compounds derived from oxidation was conducted by Escalona et al. (2014). Even at very low hydrogen peroxide doses, BPA could be degraded efficiently. At the optimal conditions (pH 3, H_2O_2/BPA of 0.20, Fe(II)/BPA = 0.012), the complete removal of up to 300 mg BPA/L was achieved in less than 2 min. During nanofiltration of BPA solution non-treated by Fenton's process, adsorption on the membrane and size exclusion were the mechanisms responsible for BPA rejection. During NF of BPA solution treated by Fenton's process, total BPA removal and over 77% in TOC, COD, colour and Fe(II) removals were obtained. NF after the Fenton's process is advantageous to reduce the catalyst (iron) concentration in wastewater before discharge and avoid the subsequent separation of iron hydroxide sludge. In addition, NF allows recycling of the soluble iron to the reaction tank for reuse during the BPA degradation, which reduces the continuous dosing of catalyst and decreases cost of treatment. Poorer permeation performance in NF of Fenton-treated effluent was related to membrane fouling by partially oxidised products.

It has been found that BPA can be degraded by the oxidation in the presence of MnO_2 (Lin et al. 2009). BPA radicals were formed via electron transfers to MnO_2. These radicals triggered reactions of radical coupling, fragmentation, substitution, and elimination, which led to the degradation of BPA. Although nanosized manganese oxides are considered promising materials for the removal of organic micropollutants because of high reactivity, environmental compatibility and cost effectiveness, their separation, recovery and reuse remain a challenge for practical applications. Several strategies have been utilized to overcome this engineering issue, including free settling, centrifugation and membrane filtration (Li et al. 2003; Pan et al. 2008b). Among these methods, membrane filtration exhibited the most excellent ability in separation of nanosized materials (Pan et al. 2008b; Zhang et al. 2009); however, fouling of the membrane modules is a major limiting factor in this application (Lee et al. 2001; Fu et al. 2006). Thus, Zhang et al. (2011) proposed a novel hybrid process combining β-MnO_2 nanowires oxidation and MF to remove BPA. It was proved that β-MnO_2 nanowires could degrade BPA effectively. BPA oxidation using β-MnO_2 nanowires was influenced by pH and humic acid and coexisting metal ions such as Ca(II), Mg(II), and Mn(II) that induced suppressive effects. Following oxidation, a crossflow MF process was conducted to efficiently separate and recover the β-MnO_2 nanowires from treated water. Almost zero turbidity of the effluent pointed out that β-MnO_2 nanowires could be easily and entirely separated by MF remaining unbroken and causing minimal pore blocking. Concluding, it can be said that the use of AOPs seems to be efficient in degrading and removing EDCs and therefore could be applied in treatment and reuse schemes where the final effluent to be reused must be of high purity. In any case though, the total dissolved solids which can potentially be increased if AOPs are applied have to be removed before any reuse application.

6.3 Technologies Combining Biological Treatment and Membrane Filtration

The combination of biological treatment of BPA-containing wastewater and membrane filtration is implemented mostly in membrane bioreactors (MBRs). As results of biological treatment and membrane filtration, Schierenbeck et al. (2003) reported the retention of dissolved organic carbon and micropollutants in the treatment system and high volumetric reaction rates at a short hydraulic retention time (HRT) of wastewater. Based on the conclusions of Göbel et al. (2007), to achieve high BPA removal efficiency, the short HRT in the bioreactor can be maintained since it will be compensated by a high concentration of biomass resulting from its retention in the system by membranes. A pilot scale wastewater treatment process used NF and RO sequenced with MBR (SRT 20 d, HRT 7.8 h) was tested for the removal of BPA at the influent concentration of 90.2 ng/L (Lee et al. 2008). BPA concentration was 6.06 ng/L in the MBR effluent, whereas polishing on membranes due to size exclusion and adsorption mechanism lowered BPA concentration to 4.08 ng/L (NF) and 3.27 ng/L (RO).

6.4 Integrated Systems Using Biocatalysts

The indigenous microbial flora of wastewater sludge can produce various enzymes, such as fungal or bacterial peroxidases, tyrosinases and laccases, which form a group of phenol-oxidising enzymes that degrade or detoxify organic pollutants. An enzyme-catalyzed polymerization and precipitation process has been widely studied as new methods for the treatment of aqueous phenols, using laccases and peroxidases. Laccases produced by *Sinorhizobium meliloti* possess the potential to degrade or detoxify micropollutants present in wastewater sludge, particularly those containing phenolic groups (Mohapatra et al. 2010).

Solid-state fermentation is the most economical process for the production of higher activity of ligninolytic enzymes by fungi. White rot fungi in wood are producers of ligninolytic enzymes which are essential for degradation of toxic organic compounds (Gassara et al. 2013). *Phanerochaete chrysosporium* is one of most studied white rot fungi, which mainly produces manganese peroxidase (MnP) and lignin peroxidase (LiP). The ligninolytic enzymes comprise LiP, MnP and laccase. Due to the complex structure of lignin, its biodegradation system is considered highly nonspecific. Ligninolytic enzymes, therefore, have attracted attention as possible degraders of structurally different environmental pollutants. The 2,2-methylenediphenol ($C_{13}H_{12}O_2$), bis(4-hydroxyphenyl)methane ($C_{13}H_{12}O_2$, bisphenol F) and *p*-(benzyloxy)phenol ($C_{13}H_{12}O_2$) have been detected in the BPA degradation pathway, which resulted in a complete BPA mineralization to water and carbon dioxide.

As the most enzymes, laccases catalyzing the transformation of micropollutant require readily available molecular oxygen as the sole cofactor to produce water as a by-product and active radicals of the substrates. These radicals generally polymerize similarly to cross-coupling mechanisms known from other enzymes catalysis and precipitate so it is easier to separate them from the reaction solution. Laccases have been shown to biotransform various aromatic micropollutants into innocuous by-products in wastewaters treatment.

The use of enzymes in wastewater treatment necessitates their insolubilization and confinement in a reactor as a part of a treatment system operating continuously. The shortcomings of free enzymes in solution are well-known. A novel approach of enzymatic treatment using a simple and efficient hybrid bioreactor combining microfiltration membrane with cross-linked laccase aggregates operated continuously was proposed (Ba et al. 2014). This system advantageously associates insolubilized biocatalysts and relatively low energy-consuming membranes, thus having a potential to become a cost-effective and sustainable technology for the polishing of wastewaters to remove micropollutants such as BPA.

The catalysts immobilized in capsules made of natural and environmentally safe materials cross-linked with glutaraldehyde showed superior stability, fast removal rate and great reproducibility. Chitosan core/alginate shell biomagnetic capsules with immobilized tyrosinase were fabricated in a layer-by-layer configuration and used for the removal of BPA and phenol (Ispas et al. 2010). The enzyme was efficiently immobilized within the biopolymer layer, which enhanced enzyme stability with no leaching. The removal capacity for BPA was slightly decreased due to the molecular size of BPA compared to phenol. The amount of BPA that can be removed per capsule is 5.6 ppm while phenol can be removed up to 10.0 ppm per capsule. The use of iron oxide magnetic particles had no effect on BPA removal, and greatly improved ease of handling. The biomagnetic capsules can be used for the treatment of phenol-contaminated waters and other aqueous solutions containing phenolics in the ppm range. Binding of the enzymatically degraded phenolics to the chitosan core produced a color change that could be used as a simple tool for visualization and confirmation of the reaction in the interior of the capsule.

To retain an immobilized enzyme with high activity, the solid supports should be biocompatible and leave the enzyme accessible to its substrates and co-substrates when used for biotransformation porous silica nanoparticles, such as fumed silica nanoparticles (Hommes et al. 2012). The production of biocatalysts should be as cost-effective as possible, in order to be implemented in wastewater treatment. The biocatalysts have to retain the ability to efficiently catalyze the transformation of BPA at environmentally relevant concentrations (i.e. concentrations in WWTP effluents). Immobilization of laccase onto fumed silica was optimized to efficiently produce industrially relevant amounts of a nanobiocatalyst for biological micropollutant elimination, whilst saving 80% of surface modification agent (3-aminopropyl-triethoxy-silane) and 90% of cross-linker (glutaraldehyde) (Hommes et al. 2012). Minimized losses during preparation and favorable effects of immobilization yielded conjugates with drastically increased enzymatic activity (164% of invested activity). Long-term stability and activity regarding BPA removal of the synthesized biocatalyst were assessed under application-relevant conditions. With 81.1 $\pm$ 0.4% of residual activity after 7 days, stability of conjugates was significantly higher than of free laccase, which showed almost no activity after 1.5 days. These data illustrate the huge potential of fumed silica nanoparticles/laccase-composites for innovative biological wastewater treatment.

Compared with traditional enzyme supports, electro-spun fibrous membranes have higher intrinsic high specific surface area and inter-fiber porosity, are easy to handle and have a low hindrance for mass transfer and good mechanical strength (Xu et al. 2013). Horseradish peroxidase (HRP) from roots of horseradish (*Amoracia rusticana*) was successfully immobilized on novel enzyme carriers, poly (methyl-methacrylate-co-ethyl- acrylate) microfibrous membranes, and used for removal of BPA from water. These membranes with fiber diameter of 300–500 nm were fabricated by electrospinning. HRP covalently immobilized on the surface of microfibers exhibited significantly higher removal efficiency for BPA in 3 h (93%) compared with free HRP (61%).

Electroenzymatic oxidation is an interesting approach that combines enzyme catalysis and electrode reactions. The immobilized HRP was used in a membraneless electrochemical reactor to catalytically oxidize BPA (Xu et al. 2011). To immobilize HRP in this reactor, the silk fibroin and poly (amido amine) were covalently bonded onto the magnetic Fe_3O_4 nanoparticles to give an aminated magnetic silk fibroin nanoparticle. The removal of BPA was monitored by applying electroenzymatic oxidation in the presence of electrogenerated H_2O_2, biochemical oxidation with externally supplied H_2O_2, and electrochemical oxidation processes. After 160 min, the BPA removal efficiency reached 80.3%. The electrochemical degradation rate for BPA was notably enhanced when the system contained immobilized HRP in the presence of electrogenerated H_2O_2. The optimum conditions for the in situ electro generation of H_2O_2 and electro-enzymatic removal of BPA were 1.6 V, pH 5.0, 25 °C and oxygen flow rate of 25 mL/min. The high degradation efficiency of 80% showed that an enzyme-based electrochemical treatment process might be a viable approach for the removal of BPA.

6.5 Production of Nanobiocatalysts for Municipal Wastewater Treatment

The production of nanomaterials has been the emerging process that may improve the removal of micropollutants from wastewater. Carbon nanotubes are hexagonal carbon lattices that are folded to form helical-like tubular structures. These structures with a large accessible external surface area have a variety of potential applications, including removal of organic micropollutants and high adsorption capacities for heavy metals (Pan and Xing 2008; Mauter and Elimelech 2008; Chen et al. 2009; Zhang et al. 2010; Perez-Aguilar et al. 2010). The adsorptive capacities and mechanisms of removal of different EDCs onto single-walled carbon nanotubes and multi-walled carbon nanotubes were investigated (Pan et al. 2008a; Oleszczuk et al. 2009), showing that they are better adsorbents than the powdered activated carbon due to higher adsorption capacities and shorter equilibrium times. One of the possible problems in carbon nanotubes utilization is their escape during the membrane filtration process and release into the environment, which may be harmful to ecosystems, leading to damage of DNA due to toxicity (Lam et al. 2006; Kostarelos 2008; Upadhyayula et al. 2009). Carbon nanotubes are still relatively expensive for large scale applications in water treatment; however, latest reports estimate that the bulk production cost of high quality carbon nanotubes becomes quite low (approximately \$10/kg) (Upadhyayula et al. 2009). The retention and adsorption of BPA was examined using three commercially available UF membranes both in the absence and presence of natural organic matter (NOM) and single-walled carbon nanotubes (Heo et al. 2012). Adsorption was an important mechanism for the retention of hydrophobic compounds. The results also suggested that BPA transport was influenced by NOM, which fouls the membrane through pore blockage and cake/gel formation. The NOM fouling was presumably attributed to the adsorptive hydrophobic interactions, which decreased the membrane pore size and caused the flux decline. A strong linear correlation between the retention and adsorption of BPA was observed, indicating that retention by the UF membranes was mainly due to the adsorption of BPA onto the membrane, the single-walled carbon nanotubes, and/or the NOM. BPA rejection showed a slight dependence on the membrane pore size.

References

Ba S, Jones JP, Cabana H (2014) Hybrid bioreactor (HBR) of hollow fiber microfilter membrane and cross-linked laccase aggregates eliminate aromatic pharmaceuticals in wastewaters. J Hazard Mater 280:662–670

Baroutian S, Gapes DJ, Sarmah AK et al (2016) Formation and degradation of valuable intermediate products during wet oxidation of municipal sludge. Bioresour Technol 205:280–285

Belgiorno V, Rizzo L, Fatta D et al (2007) Review on endocrine disrupting-emerging compounds in urban wastewater: occurrence and removal by photocatalysis and ultrasonic irradiation for wastewater reuse. Desalination 215:166–176

Bertanza G, Pedrazzani R, Dal Grande M et al (2011) Effect of biological and chemical oxidation on the removal of estrogenic compounds (NP and BPA) from wastewater: An integrated assessment procedure. Water Res 45:2473–2484

Butterfield M, Christensen PA, Curtis TP et al (1997) Water disinfection using an immobilized titanium dioxide film in a photochemical reactor with electric field enhancement. Water Res 31:675–677

Chen GC, Shan XQ, Wang YS et al (2009) Adsorption of 2,4,6-trichlorophenol by multi-walled carbon nanotubes as affected by Cu(II). Water Res 43:2409–2418

Chiang K, Lim TM, Tsen L et al (2004) Photocatalytic degradation and mineralization of bisphenol A by TiO_2 and platinized TiO_2. Appl Catal A 261:225–237

Chin SS, Lim TM, Chiang K et al (2007) Factors affecting the performance of a low-pressure submerged membrane photocatalytic reactor. Chem Eng J 130:53–63

Coleman HM, Chiang K, Amal R (2005) Effects of Ag and Pt on photocatalytic degradation of endocrine disrupting chemicals in water. Chem Eng J 113:65–72

Escalona I, Fortuny A, Stüber F et al (2014) Fenton coupled with nanofiltration for elimination of bisphenol A. Desalination 345:77–84

Fu J, Ji M, Wang Z et al (2006) A new submerged membrane photocatalysis reactor (SMPR) for fulvic acid removal using a nano-structured photocatalyst. J Hazard Mater 131:238–242

Gassara F, Brar SK, Verma M et al (2013) Bisphenol A degradation in water by ligninolytic enzymes. Chemosphere 92:1356–1360

Genc N, Yonsel S, Dagas L et al (2002) Wet oxidation: a pre-treatment procedure for sludge. Waste Manag 22:611–616

Göbel A, McArdell CS, Joss A et al (2007) Fate of sulfonamides, macrolides, and trimethoprim in different wastewater treatment technologies. Sci Total Environ 372:361–371

Heo J, Flora JRV, Her N et al (2012) Removal of bisphenol A and 17b-estradiol in single walled carbon nanotubes–ultrafiltration (SWNTs–UF) membrane systems. Sep Purif Technol 90:39–52

Hommes G, Gasser CA, Howald CBC et al (2012) Production of a robust nanobiocatalyst for municipal wastewater treatment. Bioresour Technol 115:8–15

Ispas CR, Ravalli MT, Steere A et al (2010) Multifunctional biomagnetic capsules for easy removal of phenol and bisphenol A. Water Res 44:1961–1969

Kaneco S, Rahmana MA, Suzuki T et al (2004) Optimization of solar photocatalytic degradation conditions of bisphenol A in water using titanium dioxide. J Photochem Photobiol A: Chem 163:419–424

Kim JH, Kim S, Lee CH et al (2008a) A novel nanofiltration hybrid system to control organic micro-pollutants: application of dual functional adsorbent/catalyst. Desalination 231:276–282

Kim JH, Park PK, Lee CH et al (2008b) A novel hybrid system for the removal of endocrine disrupting chemicals: nanofiltration and homogeneous catalytic oxidation. J Membr Sci 312:66–75

Kostarelos K (2008) The long and short of carbon nanotube toxicity. Nat Biotechnol 26:774–776

Lam CW, James JT, McCluskey R et al (2006) A review of carbon nanotube toxicity and assessment of potential occupational and environmental health risks. Crit Rev Toxicol 36:189–217

Lee SA, Choo KH, Lee CH et al (2001) Use of ultrafiltration membranes for the separation of TiO_2 photocatalysts in drinking water treatment. Ind Eng Chem Res 40:1712–1719

Lee JW, Cha HY, Park KY et al (2005) Operational strategies for an activated sludge process in conjunction with ozone oxidation for zero excess sludge production during winter season. Water Res 39:1199–1204

Lee J, Lee BC, Ra JS et al (2008) Comparison of the removal efficiency of endocrine disrupting compounds in pilot scale sewage treatment processes. Chemosphere 71:1582–1592

Li XZ, Liu H, Cheng LF et al (2003) Photocatalytic oxidation using a new catalyst—TiO2 microsphere—for water and wastewater treatment. Environ Sci Technol 37:3989–3994

Liang L, Zhang J, Feng P et al (2015) Occurrence of bisphenol A in surface and drinking waters and its physicochemical removal technologies. Front Environ Sci Eng 9:16–38

Lin K, Liu W, Gan J (2009) Oxidative removal of bisphenol A by manganese dioxide: efficacy, products, and pathways. Environ Sci Technol 43:3860–3864

Liu Y (2003) Chemically reduced excess sludge production in the activated sludge process. Chemosphere 50:1–7

Mauter MS, Elimelech M (2008) Environmental applications of carbon-based nanomaterials. Environ Sci Technol 42:5843–5859

Mohapatra DP, Brar SK, Tyagi RD et al (2010) Degradation of endocrine disrupting bisphenol A during pre-treatment and biotransformation of wastewater sludge. Chem Eng J 163:273–283

Mohapatra DP, Brar SK, Tyagi RD et al (2011a) Parameter optimization of ferro-sonication pre-treatment process for degradation of bisphenol A and biodegradation from wastewater sludge using response surface model. J Hazard Mater 189:100–107

Mohapatra DP, Brar SK, Tyagi RD et al (2011b) Concomitant degradation of bisphenol A during ultrasonication and Fenton oxidation and production of biofertilizer from wastewater sludge. Ultrason Sonochem 18:1018–1027

Molinari R, Borgese M, Drioli E et al (2002) Hybrid processes coupling photocatalysis and membranes for degradation of organic pollutants in water. Catal Today 75:77–85

Nakada N, Shinohara H, Murata A et al (2007) Removal of selected pharmaceuticals and personal care products (PPCPs) and endocrine-disrupting chemicals (EDCs) during sand filtration and ozonation at a municipal sewage treatment plant. Water Res 41:4373–4382

Oleszczuk P, Pan B, Xing BS (2009) Adsorption and desorption of oxytetracycline and carbamazepine by multiwalled carbon nanotubes. Environ Sci Technol 43:9167–9173

Pan B, Xing BS (2008) Adsorption mechanisms of organic chemicals on carbon nanotubes. Environ Sci Technol 42:9005–9013

Pan B, Lin DH, Mashayekhi H et al (2008a) Adsorption and hysteresis of bisphenol A and 17a-ethinyl estradiol on carbon nanomaterials. Environ Sci Technol 42:5480–5485

Pan JH, Zhang X, Du AJ et al (2008b) Self-etching reconstruction of hierarchically mesoporous F-TiO2 hollow microspherical photocatalyst for concurrent membrane water purifications. J Am Chem Soc 130:11256–11257

Papadaki M, Emery R, Abu-Hassan M et al (2004) Sonocatalytic oxidation processes for the removal of contaminants containing aromatic rings from aqueous effluents. Sep Purif Technol 34:35–42

Peng Z, Wu F, Deng N (2006) Photodegradation of bisphenol A in simulated lake water containing algae, humic acid and ferric ions. Environ Pollut 144:840–846

Perez-Aguilar NV, Munoz-Sandoval E, Diaz-Flores PE et al (2010) Adsorption of cadmium and lead onto oxidized nitrogen-doped multiwall carbon nanotubes in aqueous solution: equilibrium and kinetics. J Nanopart Res 12:467–480

Qiang Z, Nie Y, Ben W et al (2013) Degradation of endocrine-disrupting chemicals during activated sludge reduction by ozone. Chemosphere 91:366–373
Schierenbeck A, Haase C, Räbiger N (2003) Degradation of halogenated hydrocarbons by the combined application of membrane filtration with a bioreactor. Eng Life Sci 3:263–266
Strong PJ, Gapes DJ (2012) Thermal and thermo-chemical pre-treatment of four waste residues and the effect on acetic acid production and methane synthesis. Waste Manag 32:1669–1677
Sun X, Wang C, Li Y et al (2015) Treatment of phenolic wastewater by combined UF and NF/RO processes. Desalination 355:68–74
Torres RA, Abdelmalek F, Combet E et al (2006) A comparative study of ultrasonic cavitation and Fenton's reagent for bisphenol A degradation in natural waters. In: 1st European conference on environmental applications of advanced oxidation processes, e-proceedings, Chania Crete, Greece, September 2006
Torres RA, Abdelmalek F, Combet E et al (2007) A comparative study of ultrasonic cavitation and Fenton's reagent for bisphenol A degradation in deionized and natural waters. J Hazard Mater 146:546–551
Upadhyayula VKK, Deng S, Mitchell MC et al (2009) Application of carbon nanotube technology for removal of contaminants in drinking water: a review. Sci Total Environ 408:1–13
Xu J, Tang T, Zhang K et al (2011) Electroenzymatic catalyzed oxidation of bisphenol-A using HRP immobilized on magnetic silk fibroin nanoparticles. Process Biochem 46:1160–1165
Xu R, Chi C, Li F et al (2013) Immobilization of horseradish peroxidase on electrospun microfibrous membranes for biodegradation and adsorption of bisphenol A. Bioresour Technol 149:111–116
Zhang X, Pan JH, Du AJ et al (2009) Combination of one dimensional TiO_2 nanowire photocatalytic oxidation with microfiltration for water treatment. Water Res 43:1179–1186
Zhang SJ, Shao T, Bekaroglu SSK et al (2010) Adsorption of synthetic organic chemicals by carbon nanotubes: effects of background solution chemistry. Water Res 44:2067–2074
Zhang T, Zhang X, Yan X et al (2011) Removal of BPA via a hybrid process combining oxidation on β-MnO_2 nanowires with microfiltration. Colloids Surf A Physicochem Eng Asp 392:198–204

